MEGAPROJECT

CHINESE ENVIRONMENT AND DEVELOPMENT

CHINESE ENVIRONMENT AND DEVELOPMENT
A Monograph Series

Edited by Shiu-hung Luk and Joseph Whitney

With one-fifth of the world's population concentrated in about eight percent of the land surface, China's impact on her own territory throughout history has always been of a magnitude seen in few other regions. Now that China has embarked on a path of rapid economic development largely fueled by abundant resources of coal, the impact on her own land and on the globe as a whole is substantial and will increase over the coming decades. A deteriorating environment will have repercussions both on the global economy as well as on China's own progress as pollution and natural hazards intensify and increase the costs of development.

With contributions by scholars from China and abroad, this Monograph Series will explore in a holistic way, consistent with the complex ecosystem phenomena involved, the relationships between environment and development in China's past and present. This will provide valuable insights into the country's future not found in other scholarly publications.

CHINESE ENVIRONMENT AND DEVELOPMENT

MEGAPROJECT

A Case Study of China's Three Gorges Project

Edited by *SHIU-HUNG LUK and JOSEPH WHITNEY*

An East Gate Book

Routledge
Taylor & Francis Group
LONDON AND NEW YORK

An East Gate Book

First published 1993 by M.E. Sharpe

Published 2015 by Routledge
2 Park Square, Milton Park, Abingdon, Oxon OX14 4RN
711 Third Avenue, New York, NY 10017, USA

Routledge is an imprint of the Taylor & Francis Group, an informa business

Library of Congress Cataloging-in-Publication Data

Megaproject: a case study of China's Three Gorges Project /
edited by Shiu-hung Luk, Joseph Whitney.
p. cm.
"In this collection, we have selected ten papers . . . originally published in China but were translated and published by M. E. Sharpe in two special issues of the journal Chinese geography and environment, volume 1 (1988–89)"—Introd.
Includes bibliographical references.
ISBN 0-87332-733-0
1. Dams—China—Yangtze River Gorges.
2. Hydroelectric power plants—China—Yangtze River Gorges.
I. Luk, Shiu-hung.
II. Whitney, Joseph B.
III. Title: Three Gorges Project.
TC558.C52T576 1991
333.91′6215′0951212—dc20
91-22964
CIP

ISBN 13: 9780873327336 (hbk)

Contents

Acknowledgments

WE WOULD like to acknowledge the assistance of K. C. Tan of the University of Guelph and Linda Hershkovitz of the University of Toronto in editing the translated articles, preparing the bibliography in English, and providing photographs. Also we want to acknowledge the contributions of Kam-wing Chan, Weixue Cheng, Grace Chung, and Joyce Liu at various stages of this work. Finally, we thank Douglas Merwin, our editor at M. E. Sharpe, Inc., for his encouragement and support, without which the completion of this project would have been impossible.

MEGAPROJECT

Introduction

Shiu-hung Luk and Joseph Whitney

IN THE aftermath of constructing the colossal Aswan High Dam in Egypt, a host of negative environmental impacts were reported by researchers, and high dam construction around the world has been in a slump since the late 1960s. By the late 1970s, after a hiatus of a little over a decade, these huge engineering projects became fashionable again. This is especially true in developing countries. In 1986, Venezuela completed the Guri Dam, the largest in the world at the time. With a 10,000 MW generating capacity, it can produce as much electricity as ten large nuclear power plants (Shea 1988). Then there is Brazil, which is currently in the process of completing the gigantic Itaipu Dam with a generating capacity of 12,600 MW. In China, the idea of constructing a major dam on the mighty Yangtze (Changjiang) River (figure 1) dates back to the early part of this century when Dr. Sun Yat-sen, the founding father of the Chinese Republic, first suggested the feasibility of this project. Present plans under consideration in the People's Republic of China call for the construction of a megadam 1,924 m long (table 1), the creation of a gorge-type reservoir that will extend some 600 km upstream from the dam (Tao, this volume), and installation of twenty-six 500 MW turbines. When completed, the Three Gorges Dam with a proposed generating capacity of 13,000 MW will be the largest in the world.

The proposed Three Gorges Project (TGP) is a megaproject that has captured the attention of academics, environmentalists, and government institutions around the world during the last de-

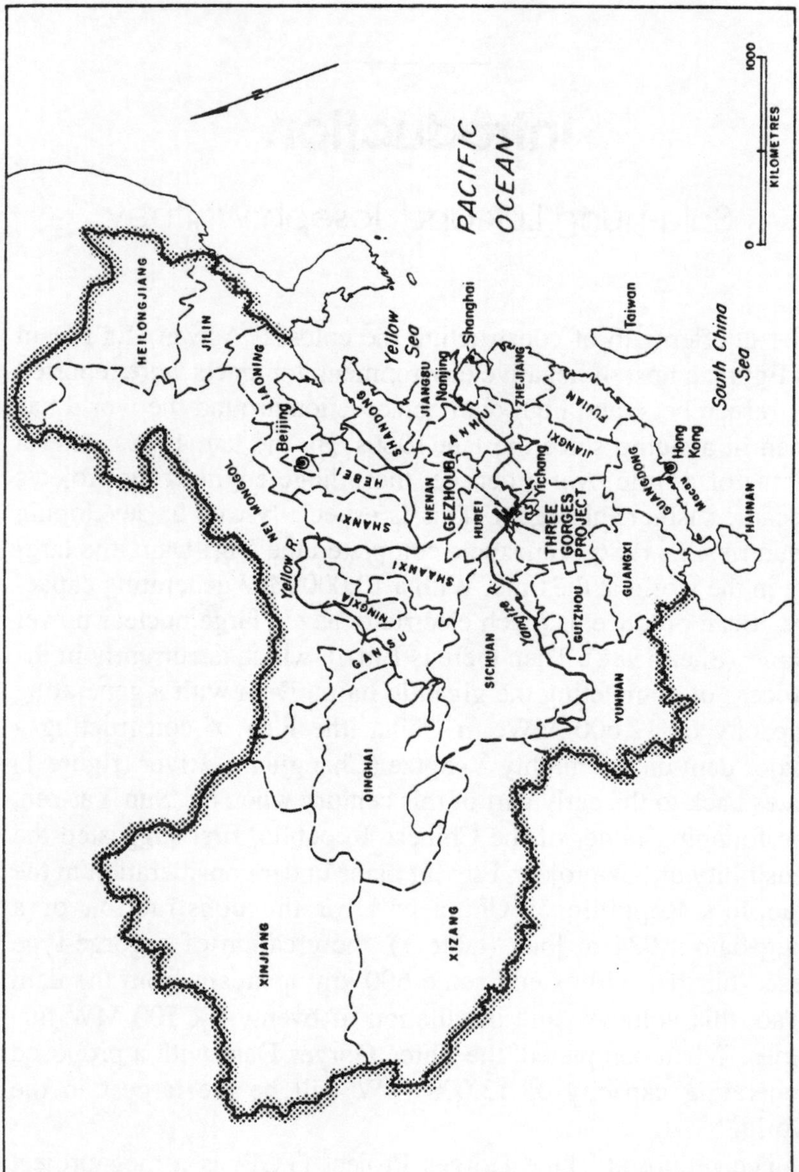
PACIFIC OCEAN
Yellow Sea
South China Sea
KILOMETRES
0
1000
HEILONGJIANG
JILIN
LIAONING
NEI MONGGOL
Beijing
HEBEI
SHANDONG
JIANGSU
Nanjing
Shanghai
ANHUI
ZHEJIANG
FUJIAN
JIANGXI
Taiwan
GUANGDONG
Hong Kong
Macau
HAINAN
GUANGXI
THREE GORGES PROJECT
Yichang
GEZHOUBA
HUBEI
HENAN
SHANXI
SHAANXI
Yellow R.
NINGXIA
GANSU
QINGHAI
SICHUAN
Yangtze R.
GUIZHOU
YUNNAN
XIZANG
XINJIANG

Figure 1.

Table 1. **Major technical and economic indicators of a range of schemes for the Three Gorges project**

Indicators		Unit	Scheme 150m	160m	170m	180m
Dam elevation		(m)	175	175	175	185
Normal pool level NPL		(m)	150	160	170	180
Low pool level during non-flood period		(m)	130	145	150	160
Flood control level		(m)	135	135	140	150
Food control capacity		($10^6 m^3$)	9,400	9,100	14,700	18,400
Flood control capacity below NPL		($10^6 m^3$)	7,300	13,800	19,700	24,900
Flood control capacity below 1000-year flood elevation		($10^6 m^3$)	22,000	22,000	19,700	24,900
20-year flood	Maximum pool level	(m)	150	150	154	156
	Maximum flood level at Shashi	(m)	44.5	44.5	44.3	44.0
	Maximum discharge at Zhijiang	(m^3/s)	56,700	56,700	56,700	51,700
100-year flood	Maximum pool level	(m)	160.7	160.7	164	175
	Maximum flood level at Shashi	(m)	44.5	44.5	44.5	44.5
	Maximum discharge at Zhijiang	(m^3/s)	56,700	56,700	56,700	56,700
1000-year flood	Maximum pool level	(m)	170	170	170	180
	Maximum flood level at Shashi	(m)	45.0	45.0	45.0	45.0
	Maximum discharge at Zhijiang	(m^3/s)	71,700	71,700	75,700	76,100
Mean regulated discharge during non-flood period		(m^3/s)	5,120	5,090	5,630	5,990
Guaranteed power output		(10^3kW)	3,320	3,810	4,600	5,370
Installed generating capacity		(10^3kW)	13,000	14,320	16,900	18,720
Annual power output		(10^6 units)	67,700	73,200	78,500	89,100
Frequency of generator use		(h)	5,200	4,940	4,640	4,760
Mean annual reduction in area flooded		(10^3 mu)[a]	470	470	457	588
Mean annual benefits from flood control		(10^6 Yuan)	610	610	590	760
Improved navigation channels		(km)	450–550	530–600	550–650	600–700
Cultivated area inundated in the reservoir		(10^3 mu)[a]	146.2	216.1	299.4	397.1
Population resettled		(10^3)	335.4	427.1	636.6	794.8
Construction works	Earth/rock excavation	($10^3 m^3$)	80,800	81,200	80,300	79,600
	Earth/rock fill	($10^3 m^3$)	31,200	31,200	31,200	91,100
	Cement	($10^3 m^3$)	22,700	22,900	23,300	25,900
	Reinforcement steel	(10^3 tonnes)	247	250	252	276
	Structure steel	(10^3 tonnes)	227	235	235	255

Source: CAS 1988: 14.

[a]1 mu = 0.07 hectare.

cade. In China, the proposal was given serious consideration in the 1950s but proved to be controversial even in those early days. In 1958, the late Premier Zhou Enlai called a meeting in Chengdu of the leading members of the central government during which the long-term objective of building the giant dam was confirmed but it was decided that important conflicts had to be resolved before the project could proceed. During the ensuing decade, the proponents of the project, especially the Changjiang River Basin Planning Office (Changjiang Liuyu Guihua Bangongshi, POCRB) lobbied hard, but with the deteriorating economic conditions, the opportunities to build the TGP were fading (Lieberthal and Oksenberg 1988: 307). By 1970, perhaps to relieve the mounting political pressure, the central government approved the Gezhouba (GZB) Dam project, which was located some 40 km downstream from the current proposed site of Sandouping for the TGP, and which was planned to have about one-fifth the generating capacity of the TGP. It was supposed to be a warm-up project before the giant TGP was started (Li 1985, Hong, this volume). As far as we know, no detailed feasibility study of the GZB project was conducted. In the proposal of the GZB project, a total investment of 1.35 billion yuan (current exchange rate is approximately U.S. $1 = ¥5.5) and a construction period of five years were estimated. In the end, it cost 5 billion yuan and took nineteen years to complete. The GZB was officially opened in 1989.

With the adoption of an "open door" policy and an emphasis on economic construction in 1979, the POCRB again pushed for the construction of the TGP (Hong, this volume). Presumably as a consequence of the negative experiences in the development of the GZB, the leading members of the central government took a much more cautious approach. Following the recommendation of members of the United States Bureau of Reclamation who visited the area in 1981, the government commissioned detailed feasibility studies of the TGP. During 1984–85, the feasibility study in China was coordinated by the Science and Technology Commission (Kexue Jixu Weiyuanhui), which was directly responsible to the State Council (Zhongguo Guowuyuan). For reasons not entirely known to us, however, the Ministry of Water Conservancy and Electric Power (Suili Dianli Bu) took over the coordination work in 1985. As a result, among the fourteen consultation study groups organized, ten of them were headed by members of the Ministry of Water Conservancy and Electric Power and the POCRB, which were known proponents of the project. In 1986, the World Bank and the Canadian International Development Agency commissioned a Canadian consortium (CIPM Yangtze Joint Venture) to conduct a $14 million feasibility study of the same project (CYJV 1988). The final reports prepared by both the Chinese and the Canadian team were submitted to the Chinese government in 1988.

During 1987–89, many articles and research papers on the feasibility of the TGP megaproject were published in China. They include several major volumes prepared by members of the feasibility study team in the Chinese Academy of Sciences (Zhongguo Kexueyuan) (CAS 1987, 1988). In addition, many articles published in magazines and newspapers expressed the opinion that views not favoring the immediate construction of the megaproject were underrepresented in official reports and publications, implying that a democratic process to collective decision making was not followed. Hence, several collected essays (e.g., Tian and Lin 1988, 1989) were published to provide a forum for such dissenting views. Other, shorter collections have also been distributed (Dai 1989).

In this collection, we have selected ten papers that provide arguments for and against the construction of the TGP. These papers were originally published in China but were translated and published by M. E. Sharpe in two special issues of the journal *Chinese Geography and Environment*, volume 1 (1988–89). In selecting these papers, our objective was to review the major issues related to the Three Gorges Project that have been raised by various Chinese authors and other researchers and consultants. To enliven the discussion on the major issues and to introduce the selected articles, we have prepared this short essay. We hope this will provide the necessary framework by which pertinent questions about the TGP can be raised. In addition, a bibliography of TGP sources published in English is appended.

The TGP: Major controversies

Up and downstream flooding

One of the major objectives of constructing the TGP is to reduce the flood hazard downstream of the dam by impounding water behind the proposed dam. As was indicated in many of the articles translated, flooding has always been a serious problem in the middle reaches of the Changjiang, but especially in the Jingjiang section located between Jiangning and Jianli, and also at Wuhan, which is affected by flood waters from the Jingjiang section, the Hanjiang River, a tributary of the Changjiang, and other rivers of the Dongtinghu Lake system (figure 2). Currently, the Jingjiang section is protected by embankments that can withstand a flow not exceeding 60,000 m^3/s. During the last one hundred years, there were twenty-one floods that exceeded this discharge at Yichang, and thus serious flooding occurred in about one in five years.

Since the 1950s, the excess flood water was diverted via sluices, which include the Jingjiang sluice located near Shashi (which has a capacity discharge of 8,000 m^3/s) and the Dujiatai flood diversion structure on the Hanjiang River (which can handle a discharge of 6,000 m^3/s) (Kennedy 1987). With diversion, flood discharges of up to about 70,000 m^3/s can be withstood. It should be mentioned, however, that the diversions have required emergency evacuation and inundation. In fact, as much as 710 km^2 of flood diversion area and a population of about 3.1 million can be affected (CYJV 1988: 7:1–2). During the 1954 flood, which was induced by basinwide rainstorms, a maximum discharge of 76,100 m^3/s was recorded at Wuhan (CAS 1981). While serious damages to the embankments were averted, one report suggests that as many as 30,000 people perished (Sun et al.

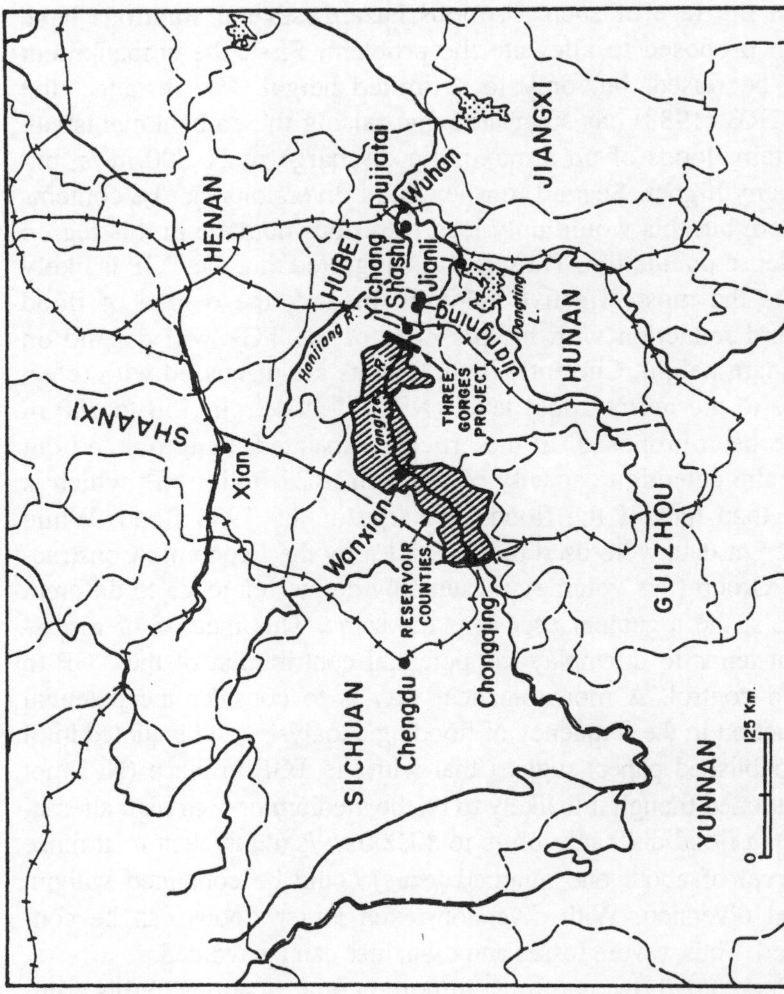
SHANXI
HENAN
HUBEI
JIANGXI
HUNAN
GUIZHOU
SICHUAN
YUNNAN
Xian
Wanxian
Chengdu
Chongqing
RESERVOIR COUNTIES
THREE GORGES PROJECT
Yichang
Shashi
Jianli
Wuhan
Dujiatai
Jingjiang
Dongting L.
Hanjiang R.
Yangtze R.
0
125 Km

Figure 2.

1987). The estimated flood discharge of the 1870 event, the highest on record, was 110,000 m^3/s at Yichang (POCRB 1979). The colossal flood caused immense damage and many casualties.

In the face of such a serious hazard, several solutions have been proposed to alleviate the problem. First, the embankment can be raised, but only to a limited height. For instance, the POCRB (1988) has suggested that raising the embankments can contain floods of up a maximum discharge of 80,000 m^3/s, but not any higher. Second, further flood diversions can be contemplated, but this would only lead to serious damage in this region of dense population. Thus, it was suggested that the TGP is likely to be the most effective solution. In part, the degree of flood hazard reduction with the presence of the TGP will depend on the dam height. Currently, dam heights are discussed with reference to the normal pool level (NPL). NPLs from 150 to 180 m have been proposed. In the articles translated, Fang pointed out that the detention capacity of the dam is 22 billion m^3, which is less than half of the flood discharge of the 1954 flood. While different data were used in the articles by the Economic Construction Group (this volume), presumably due to reference to different NPL's, the argument used was the same. This method of assessment tends to downplay the potential contribution of the TGP to flood control. A more practical way is to consider the potential reduction in the frequency of flooding. Analysis results quoted in Si (unpublished paper) suggest that with the TGP in place (NPL not mentioned though it is likely to be the medium or high dam alternative), a flood discharge of up to 80,000 m^3/s (equivalent recurrence interval of about one hundred years) could be contained without flood diversion. With diversion, even larger floods can be contained. Thus, severe losses and casualties can be avoided.

There are, however, three important qualifications to this argument. First, it is well known in the experience of flood control that no matter how large the dam is, it can never stop downstream flooding completely. So while flood damage will be reduced, damages in the downstream locations due to maximum possible floods that occur as a result of prolonged and basinwide rainfall will continue to be inflicted. Second, the cost of flood protection increases sharply if the area is to be protected from infrequent floods. In the articles translated, there are references to the disastrous 1870 flood, which caused tremendous damage, and suggestions that with the TGP in place, a huge reduction in damages can be obtained even if the 1870 flood were to recur. It should be borne in mind that this is an event that has a return interval estimated to exceed one thousand years, and it is questionable whether is is economically viable or even desirable to design a project that will prevent losses from floods of such magnitude that will only occur, on average, once every one thousand years. Third, flood reduction benefits vary with the magnitude of the event, and substantial benefits during a major event that is infrequent must be balanced against lower benefits during low to medium events that occur more frequently. For instance, according to the CYJV report, of the total flood reduction benefits, 70 percent of the "expected" (in a statistical sense) benefits go to major diversion areas and beach areas that are unprotected by the dikes (CYJV 1988: 7:9–10). It is when the recurrence of the 1870 flood is considered that 81 percent of the benefits go to the protected areas (i.e., major population centers such as Wuhan). Hence, this lends support to the idea that the project provides major benefits to the protected areas only when the thousand-year flood occurs. Most other times, it benefits the diversion areas where potential damages could have been reduced significantly by floodproofing methods. Given this distribution of benefits, it is worthwhile to consider alternatives or at least auxiliary measures to reduce damages in the major diversion areas.

One of the consequences of constructing the TGP is inundation in the reservoir area as well as flooding by the backwaters during the detention period. The problems related to the loss of valuable productive land in the valley will be discussed later in this essay under the subject of resettlement. Here, we will focus on the question of flooding in the areas near Chongqing. While it was claimed that a large amount of research on the physical and mathematical modeling of the flood level at Chongqing was conducted (Qian et al., this volume), the answers provided by the different sources are divergent. Qian et al. suggested that this level will be 195 m, 196.5 m, or 198 m for NPL 140–150 m. Fang (this volume), however, claimed that the calculations are unreliable and that the maximum water level will reach 210 m and 216 m when water discharge reaches 66,500 m^3/s and 110,000 m^3/s respectively. No reference is made to NPL. CPPCC (1986) referred to data published by the Chinese Academy of Sciences that for NPL 180 m, the water level at Chongqing will reach 205 m during a hundred-year flood.

This divergence of opinion is not easy to comprehend. The authors of the CYJV report recognized that current predictions of the upstream flood heights are not reliable. In part, we believe that this is the result of the use of different assumptions. One of the key parameters for estimating the water surface gradient is channel roughness, and Qian et al. have discussed this issue in some detail. Another major question here is, of course, the mode of operation of the dam, which determines the normal low water level during the flood season. It appears to us that the final answer will largely depend on the caliber, rate, and location of the sediment that will be deposited behind the dam (because roughness is closely related to particle size of the sediment). Without further information, we can only make the following suggestions. Keeping a low water level will likely minimize upstream flooding, but it will also reduce the capacity for power generation. Even when water level is kept at the lowest possible level, it is within the realm of probability that some inundation in the Chongqing area will occur at extremely high flows, and this should be accounted for in the cost-benefit analysis.

In the TGP literature, many Chinese authors have posed the question of whether it is justifiable to uproot a million people and flood huge areas in the TGP reservoir in order to protect downstream areas from flooding. In this regard, two major arguments have been presented by the opponents of the project. First, flood hazard is also very high in the upstream areas in the Sichuan Province (as shown by a major flood in 1981), but somehow the

POCRB gives this a very low priority. In fact, the POCRB typically emphasized the high flood risk of the downstream areas by referring to the magnitude of the thousand-year flood, but the standard for flood protection upstream (e.g., for Chongqing) was set at the hundred-year flood level (Fang 1989b: 315). Second, the actual flood hazard may have been exaggerated. For instance, the records for the 1870 flood, which was the largest in history, are fragmentary. Determination of flood hazard up to the one-thousand-year flood on the basis of such records is unreliable (Lu 1989: 302). In addition, the POCRB had suggested that if the 1870 flood were to recur without the TGP dam, the Jingjiang embankments will breach on both the north and the south side, and 500,000 casualties may be inflicted. In reality, the 1870 flood did not cause major breaches of the embankment on the north side, and the embankments have since been strengthened substantially. Hence, the suggested level of casualties is unfounded (Lu 1989: 303).

Upstream land use changes and reservoir sedimentation

As many Chinese authors suggested, sedimentation in the reservoir is a major concern, particularly in view of the serious sedimentation problem experienced at the Sanmenxia Reservoir in the Huanghe (Yellow) River. On this issue, Qian et al. argued that by adopting the "impounding the clear water and discharging the turbid water" mode of operation, the rate of reservoir sedimentation will not be excessive, and much of the sediment will be flushed out during the high flow period. This is based, however, on the assumption that the current rate of land erosion and sediment transport in the river would not increase. Review of existing information suggests that this assumption may not be valid.

During the last decade, several Chinese sources have emphasized the recent trends of land use changes in Sichuan Province, and the increasing rate of sediment input into the Changjiang River. For instance, it was suggested in CAS (1988: 45) that the total sediment load at Yichang has a long-term mean of 514 million tonnes, and since 1978 there were some increases in the sediment load. In 1980 and 1984, it reached 729 and 672 million tonnes, an increase of 42 percent and 31 percent respectively over the long-term mean. Considering the many reports on mounting population pressure and land degradation in the upstream areas (table 2) (Shi et al. 1987, Shi 1989, Fang 1989a, He 1989, Wang, this volume), we can expect a trend of increasing input of sediment into the Changjiang River. However, the fact that the current rates of sediment transport in the Changjiang at Yichang are not as high as might be expected can be attributed to the trapping of sediment by various water conservancy projects (CAS 1988: 163) (tables 3 and 4). As the upstream reservoirs are gradually filled up with sediment, these effects are likely to be reduced in the near future, and large increases in the sediment load can be expected.

In this regard, what is often deceptive in the comparison between the Changjiang and the Huanghe rivers is the difference in sediment delivery (He 1989). In most parts of the Huanghe River Basin, the sediment delivery ratio approaches 1.0, which suggests that virtually all the eroded material from the hillsides will be transported by the river. In the Changjiang, however, the delivery ratio is much lower because the sediment is considerably coarser. In fact, less than 20 percent of the eroded sediment is directly transported by the river, and the bulk of the eroded sediment is deposited on floodplains and other low-lying areas, ready to be flushed by future floodflows. Consequently, just considering the actual sediment mass transported per year is inadequate in evaluating sediment movement in the basin (CAS 1988: 45).

After the complete siltation of the sediment pool behind the TGP dam, which would likely take place in a period of seventy to eighty years, Qian et al. suggested that an equilibrium would be reached and the selected mode of operation of the reservoir will prevent further sedimentation. This is certainly true for suspended load, or the finer sediments, but coarser sediments that are usually transported as bedload will continue to deposit. In

Table 2

Changes in area affected by soil erosion in the Changjiang River Basin, 1957–1980

	1957		Early 1980s		1957–1980s	
Province	Area affected (km^2)	Percent of total area	Area affected (km^2)	Percent of total area	Increase in area (km^2)	Percentage increase
Jiangsu	1,850	3.8	6,100	12.3	4,250	229.0
Anhui	13,686	21.3	19,263	30.0	5,577	40.7
Jiangxi	11,000	6.6	38,360	23.0	27,360	248.7
Hunan	55,880	27.6	56,640	27.9	760	1.4
Sichuan	93,380	16.1	382,000	67.3	288,620	309.1
Guizhou	12,816	11.3	35,300	31.2	22,484	175.4
Total basin	363,790	20.2	739,376	41.0	375,586	103.2

Source: CAS 1988:4.

Table 3

Completed water conservancy projects in the TGP catchment area, Changjiang River

River system	Large structures		Medium size structures		Small structures type A		Small structures type B		Ponds		Total	
	Capacity ($10^6 m^3$)	Percent of total	Capacity ($10^6 m^3$)	Percent of total	Capacity ($10^6 m^3$)	Percent of total	Capacity ($10^6 m^3$)	Percent of total	Capacity ($10^6 m^3$)	Percent of total	Capacity ($10^6 m^3$)	Percent of total
Peijiang River	294	7.3	577.39	16.4	405.90	11.0	199.68	13.2	423.12	10.7	1,900.09	11.4
Jialingjiang River	1,789	44.1	256.07	7.3	243.52	6.6	192.08	12.7	470.66	11.9	2,951.33	17.6
Qujiang River			462.69	13.1	161.87	4.4	140.89	9.3	783.01	19.3	1,548.46	9.3
Minjiang River	675	16.6	241.70	6.9	341.44	9.3	104.55	6.9	374.73	9.5	1,377.42	10.4
Tuojiang River	185	4.6	542.88	15.4	662.23	18.0	285.69	19.0	873.24	22.1	2,549.04	15.2
Changjiang main Stem (upper reaches)	1,112	27.4	312.52	8.9	811.49	22.0	464.05	30.8	953.46	24.1	3,653.52	21.8
Jinshajiang River			953.03	27.9	687.26	18.7	105.37	7.1	55.92	1.4	1,801.58	10.8
Wujiang River			175.75	5.0	366.45	10.0	14.66	1.0	21.14	0.5	578.00	3.5
Total capacity	4,055		3,522.03		3,680.16		1,506.97		3,955.28		16,719.44	
Total number of structures	8		153		1,568							

Source: Gu et al. 1987 : 538.

Table 4

Sedimentation in selected reservoirs, Changjiang River Basin

Type of reservoir	Number of reservoirs	Reservoir capacity ($10^6 m^3$)	Reservoir sediment ($10^6 m^3$)	Reservoir capacity lost to sedimentation (%)
Large reservoirs	23	25,638	1,293	5.04
Medium size reservoirs	12	257.546	24.181	9.09
Small reservoirs	182	92.343	19.7825	21.42
Ponds	6492	74.24	41.48	55.90
Total	6709	26,062.129	1,378.4435	5.29

Source: CAS 1988: 4.

fact, existing data show that bedload has been increasing in the Three Gorges section of the Changjiang River (He 1989: 321). Also, at the present time, bedload only constitutes about 2 percent of the total sediment load at Yichang (CYJV 1988: 5). After the sediment pool is completely filled, it can be expected that about 6 percent of the total sediment will be bedload (He 1989: 322). With lowered gradient in the sediment pool as sedimentation continues, the rate of deposition of coarse sediment may increase further, which will significantly shorten the economic life of the reservoir. Previous reports by the POCRB (Gu et al. 1987) suggest that dam construction upstream of the Three Gorges is required to reduce substantially sediment input to the Three Gorges reservoir. Technically, this will be effective in reducing sedimentation at the Three Gorges reservoir, but the economic viability of this proposal has to be demonstrated.

A major problem of sediment management in this megaproject is deposition in the backwater zone located near Chongqing, which will aggravate the related flood hazard as well. Qian et al. suggested that the impact of sedimentation and flooding will be limited. They emphasized that retrogressive scouring following the lowering of the water level at the dam toward the end of the dry season would help to remove sediment accumulating near the backwater zone during high water levels. However, only NPL 150 m was considered in their study. Other sources from China indicate that Chongqing could be severely affected by sedimentation (Sun 1989, Fang, this volume, 1989a). Due to the difficulty of providing reliable estimates of the flood heights at Chongqing, precise locations for sedimentation in the port of Chongqing are difficult to predict. Zeng (1989) suggested that in the Huanghe River, experiences of adopting the mode of operation of impounding the clear water and discharging the turbid water were quite successful in removing accumulated sediment in the lower part of the Sanmenxia Reservoir following reconstruction of the Sanmenxia project in the early 1970s. Retrogressive scouring, however, was not significant 120 km upstream. In the case of Chongqing, which is located over 500 km from the Three Gorges dam, it is doubtful whether sediment deposited in Chongqing will be scoured, especially with the presence of the narrow Tongluoxia Canyon, which is located only 15 km downstream of Chongqing and which will serve as a bottleneck controlling the processes of channel scouring upstream. Clearly, this problem is unresolved. Assuming that significant sedimentation occurs in the port of Chongqing, frequent dredging will be required, and certainly, we need to know the cost of dredging and the impact of the sedimentation on navigation.

Scouring below the dam

The issue of scouring below the dam is one of the major criticisms leveled by Fang. Recent experiences in large dams in other parts of the world have shown that this will likely occur. In the case of the Jingjiang section, it has been reported that the embankments are beset by numerous problems, including a rather unstable sand and gravel foundation, frequent cave-ins, deep pools of water built up behind the embankment, and levees weakened by ants. Thus, frequent maintenance is required (Kennedy 1987). At present, a program of raising the embankments by 1–2 m is ongoing. Hence, there seems to be little question that the embankments below the dam do require special protection. In the CYJV report, it was suggested that some degradation in the channels below the dam is unavoidable, and that these problems can be managed by increasing low flows through regulation, river training works, and some dredging (CYJV 1988 5:1–10). While these measures are technically feasible, it is important to determine how the increase in low flows may be in conflict with other objectives, such as power generation. In addition, the costs involved should be properly estimated and included in the project cost-benefit analysis.

Impact on the delta area

An issue seldom mentioned is the fact that because of the filtering effect of the dam, the size of sediment transported by the

Changjiang River will be reduced, and this may induce increased flocculation of the fine sediment at the delta area when they come into contact with sea water. Under specific circumstances this will produce a layer of "fluidized mud," which will enhance the level of flood water in the Shanghai region (Fang, this volume). Ren (1987) considers this phenomenon unlikely, based on the observations that the distribution of fluidized mud is limited in area, that the volume of accumulation at the estuary is not increasing, and that conditions at the estuary are not favorable for the formation of this fluidized bed.

A question of perhaps greater significance is the potential problem of sea water incursion and deterioration of water quality. Wang argued that this is only of minimal significance. Qian et al. suggest that the water quality in the estuary will not be significantly affected because the Three Gorges dam has a very small capacity-inflow ratio, and it would have a minimal effect on the river flow except for its seasonal regulation. This depends on the mode of operation of the dam, however, because the timing of water release and the regulated flow rate will significantly affect the extent of sea water incursion. Si (1987) suggests that the quality of domestic water supply may be seriously affected in view of the dynamics of streamflow-tide interaction.

At present, sea water incursion is most prevalent during the period of strong tide (October to January), but the process is reversed during the period of weak tide (February to April). With the construction of the TGP, the tidal effect on salt water incursion will be enhanced due to decreased regulated flow in the strong tide period. During the weak tide period, the increased regulated flow will maintain a higher groundwater table, thus reducing the effectiveness of the normal flushing of salts in the soil profile.

Furthermore, research conducted by CAS (1988: 102) indicates that with regulated flows, daily discharge will be increased during January to April, after the reservoir has been impounded. There will be a significant decrease in discharge during October, however, when the reservoir is filling. The net effect for the whole year is for the frequency of sea water incursion to increase. The actual amount of increase would depend on the level of regulated flow. Hence, to reduce this effect, a minimum regulated flow of 18,000 to 23,000 m^3/s at the Datong station is recommended (Wang Chaojun, this volume). The question here is to what extent this will interfere with the process of reservoir impoundment in October.

Resettlement

According to the CYJV report, depending on the dam height selected, the area inundated by the TGP varies from 335 km^2 for NPL 150 m to 737 km^2 for NPL 180 m. In proportion to the total farmland in the nineteen counties located within the reservoir area (table 5), farmland lost to inundation represents 1.6 percent at NPL 150 m and 4 percent at NPL 180 m. The population to be resettled ranges from 539,000 for NPL 150 m to 1.2 million for NPL 180 m. In addition, ten to thirteen county seats and four hundred to six hundred industries will be displaced. Similar figures were cited by other authors (Wang; Tian and Lin, this volume). While this type of resettlement is not new, the scale involved surpasses that of major reservoirs constructed in China at Sanmenxia, Danjiangkou, Wujiangdu, and so forth. Given the numerous problems, many of which are still unresolved in these other projects, it is questionable whether the lot of the displaced population will be any better this time. Moreover, not only is the life and livelihood of thousands of resettled families involved, but there are hardships imposed on the host population already living in the area targeted for resettlement, and a situation of conflict and extreme stress between the two groups is always present. The Western reader should be careful not to judge the resettlement of hundreds of thousands of people in the same way as in the West. There, such a displacement would be of major social and political concern as to prevent a scheme of this magnitude from being undertaken. Criteria other than those used in the West, however, are employed at present to assess the political feasibility of resettlement in China.

Table 5

Basic statistics for counties and cities in the Three Gorges reservoir area, 1984

County/Shi (City)	Administrative area (km^2)	Cultivated area (km^2)	Population (10^3 persons)			Total industrial and agricultural production (10^3 yuan) Yichang
			Urban population	Agricultural population	Total population	
Zigui	2,427	43.68	28.7	371.3	400.0	149,210
Xingshan	2,328	23.47	17.9	159.5	177.4	90,560
Badong	3,580	68.07	26.4	429.1	455.5	146,960
Wushan	2,954	65.14	26.6	480.3	506.9	152,080
Wuxi	4,000	65.78	30.2	434.9	465.1	147,920
Fengjie	4,099	91.98	62.2	799.0	861.2	218,320
Yunyang	3,634	103.25	74.1	1,038.5	1,112.6	299,500
Kaixian	3,956	121.01	75.4	1,224.5	1,299.9	459,210
Wanxian	3,220	99.31	68.5	1,125.8	1,194.3	364,880
Wanxian Shi	241	9.43	134.1	142.4	276.5	442,800
Zhongxian	2,177	84.58	51.7	802.5	854.2	282,500
Shizhu	2,887	44.75	29.0	396.7	425.7	122,340
Fengdu	3,801	81.77	46.9	669.1	716.0	242,740
Peiling	2,824	110.10	138.5	826.3	964.8	465,250
Changshou	1,443	73.50	88.3	722.4	810.7	167,450
Wulong	3,008	50.04	28.5	335.4	363.9	126,000
Baxian	3,551	130.40	105.8	1,072.5	1,178.3	458,730
Jiangbei	2,167	91.10	74.3	906.1	980.4	335,310
Chongqing Shi (urban area only)	1,534	50.80	2,072.0	702.0	2,774.0	850,690
Total	57,752	1,469.49	3,226.9	13,132.3	16,359.2	5,889,660

Source: CAS 1988:12.

Nevertheless, it is disturbing that none of the writings reviewed, either Chinese or foreign, present any evidence that the people affected by the scheme, as opposed to local authorities, have been or ought to have been consulted in any way about the impact the move will have on their lives. In fact, perhaps with a few exceptions, virtually none of the authors opposed to the TGP cited population displacement as a major argument to support their opposition to the project. In the TGP proposal, it was suggested that the displaced population can migrate upslope, so that even though they have to abandon their hometown or village, they can remain in the same district (Tian and Lin, this volume: 187). On paper, this may seem feasible, but when the conditions in the upland areas are examined, it becomes obvious that this scheme is not likely to be successful. Fundamentally, the problem is that the best land in the reservoir area is located in the valleys that will largely be inundated, and the remaining land upslope is of much inferior quality. Apart from the steeper slopes, the soils in these upland areas are infertile, and productivity is many times lower under natural conditions. Currently, this upland area has a per capita cultivated land of 0.07 ha, and per capita grain production of 340.5 kg. Relative to the national average of grain production, Chen (1987) suggested that this area is already 15 percent overpopulated. Thus, increasing the population density is not viable. Moreover, although the CYJV report indicates that the amount of "claimable" land far exceeds that inundated (CYJV 1988: 9:11–19), the aerial survey upon which this assessment was made was confined to only forty-three "typical" xiang (rural districts) out of a total of six hundred. As the CYJV report admits (CYJV 1988: 9:7–7), there are discrepancies in the definition of "claimable" in the Chinese reports, and it is by no means clear how the "typical" counties were selected. A proper method would have been to have chosen the xiang in a random fashion, since there is always the chance that unconscious biases may be present when "typical" examples are selected.

The project proposal emphasized integrating resettlement with resource exploitation (Tian and Lin this volume: 190), but the availability of capital for development and suitable resources for exploitation are questioned (Chen 1987). For instance, the development of salt mines may have to face stiff competition from well-established facilities in Sichuan, and the development of tourist industries will be in conflict with the establishment of certain types of industries that will cause serious pollution problems.

One important aspect of resettlement is the impact on land degradation in the resettled areas (Wang, this volume). Detailed research on this topic has been conducted by Professor Shi and his colleagues at the Nanjing Institute of Pedology (Shi et al. 1987). It was suggested that to replace the 400,000 *mu* of prime agricultural land inundated, new land up to 2,000,000 mu had to be provided. This is because of the lower productivity of the upslope areas. The clearing of new land and demand for firewood for fuel will lead to an additional sediment production of 4 to 7 million tonnes per annum. In addition, other activities, including mining and extraction of building material, will yield at least another 5 million tonnes. Of this total erosion of 10 to 12 million tonnes, assuming a sediment delivery ratio of 0.5, it will add some 6 million tons of sediment per annum to be deposited in the reservoir, an increase of 15 percent over current rates. The actual increase may be much higher. Yet in the official sources from China and in the CYJV report, there is virtually no consideration of this problem. In fact, official Chinese sources tend to neglect the effect of human activities on land degradation (Fang 1989a). Obviously, there are technologies that can be used to reduce erosion, but their costs tend to be too high, and thus it is unlikely to be economically viable. Hence, with the increased population pressure and the need for infrastructure development (road building, etc.), we can expect an enhanced rate of soil erosion that will further increase the input of sediment to the reservoir. Clearly, this issue has to be examined to reveal what is a realistic expectation of sediment input from the area around the reservoir.

According to the CYJV report, after the construction costs of the dam, resettlement costs come second at 34 percent of the entire project for NPL 160 m, the alternative favored by the

CIPM consortium. As high as these values are, it is not apparent that they cover the full costs of relocation. The following considerations would show why this is uncertain.

1. According to the CYJV report (1988 9:7–10), one-half of the claimable land is situated at elevations above 800 m where development is potentially more expensive due to the cost of access and the fact that cultivation is limited to a smaller range of crops than at lower elevations. It is not clear that these additional costs have been included in the estimates.

2. It is believed that as much as 30 percent of the urban population residing in the inundated area are so-called floating population, or illegal residents who have come to the cities from the countryside in defiance of government regulations restricting such moves. The government does not want to "reward" these illegal residents by bearing the costs of their resettlement (Fearnside 1988), which have not been included in the cost estimates.

3. The additional costs of erosion control brought about by the conditions described above have not been included.

4. The cost of relocating cities and towns has almost certainly been underestimated. The CYJV report estimates (1988: 1:17–19) that the cost of relocating urban areas is about 3.1 billion yuan, or approximately the same per unit area cost as the rebuilding of Tangshan, a northeastern city totally devastated by an earthquake in 1976. However, As Fang and Wang pointed out (this volume: 180), the latter city was built on level terrain compared to the rugged topography in which nearly a dozen county seats and scores of towns will have to be relocated together with their infrastructure of roads, water supply, and so forth.

Hydroelectric power

The three basic reasons for constructing the TGP are flood control, improvement of navigation, and generation of electricity. The first objective prevents flood damage but does not generate income; the second may bring in some revenue derived from the passage of vessels through locks and will certainly generate revenue, indirectly, by stimulating the economies of regions made more accessible by improved navigation. Only the third objective, hydropower, will generate direct revenue that can be used to finance the gigantic loans necessary for the construction of the TGP. Generation of more electric power is also a top priority to achieve the four modernizations. Li (1985) estimates that currently China has a shortfall of electricity production amounting to some 40–50 billion kWh per year, approximately the annual output of the proposed TGP generators, and that this shortfall is increasing rapidly. This is especially true in the Central and Lower Changjiang valleys, where Chen (1986) estimates that by the year 2000 the supply of electricity will meet only 20–30 percent of the demand at the present rate of new generation capacity. These shortfalls have a damaging effect on the growth of the economy. If each kilowatt-hour creates an industrial output value of 2.8 yuan, then the present shortfall of 50 billion kWh represents a loss in output value of around 140 billion yuan and a loss in taxes and profits of about 30 billion yuan (Li 1985).

Four basic questions, therefore, need to be answered:

1. Will the 13 million kW of proposed generation capacity in the TGP actually be able to generate the 60 billion kWh of electricity according to the operating schedule of the dam?

2. Will the predicted revenue from electricity of 18 billion yuan during the seventeen years of construction (Wang Jiazhu, this volume: 200) be achieved?

3. Will the necessary transmission lines, transformers, and adjunct thermal generators be in place so that the electricity generated at the TGP can be effectively used?

4. Has the cost-effectiveness of alternative sources of power, such as nuclear and thermal, or alternatives to the one large dam at the TGP been adequately assessed (Tian et al., this volume: 209–210)?

In this section, we shall consider the first three questions and leave the fourth to the section dealing with financial and economic considerations.

The total annual power produced by a hydroelectric plant depends on (1) the size of the installed generating capacity; (2) the head of water in the reservoir summed throughout the year; and (3) the mass of water passing through the turbines. If the turbines were operating at maximum capacity throughout the year, the annual power generated would amount to $13 * 10^6$ kW * 8,760 hours = 113 billion kWh. The actual power predicted to be generated amounts to only 60 billion kWh, however, which means that the capacity utilization rate (energy supplied/energy that could be supplied if the generating plants were fully utilized on a year-round basis) is only 0.53, that is, 53 percent of the potential generating capacity will be used. There are two main reasons for this: turbines may be closed for part of the year for maintenance; and the head of water and flow may be insufficient to drive the turbines at their maximum output.

Clearly, the elevation of the head of water is the factor most subject to variation. As mentioned, the operating principle of the TGP is to impound clear water (i.e., at low-flow periods) and discharge turbid water (i.e., at flood periods). This means that the reservoir is kept at a low level of 135 m during the flood season from June to September (Qian et al., this volume: 135). During this period the head will be low and minimum power will be generated. In October, impoundment begins, and by the end of November, the normal pool level of 150 m will be attained. At this time the generators will be functioning at maximum capacity. During the dry season (January–April), the pool level is gradually lowered to its minimum level of 130 m (Qian et al.: 135). Hence only during one or two months will the head of water be sufficient to obtain the design output of the generators; for the remaining ten months, power will be produced at submaximal levels. In light of this, it appears that the projected 60 billion kWh to be produced each year may be overoptimistic.

It should also be emphasized that whatever annual power generation is actually achieved, this level will not be reached until the dam is completed, seventeen years after construction begins, or probably well into the first decade of the twenty-first century, by which time all twenty-six turbines will be installed. Prior to that time, however, there will be a phased introduction of turbines (Wang Jiazhu, this volume: 200), so that by year eleven, two groups of turbines with 1 million kW of generating capacity will be installed. In each subsequent year until the seventeenth, four more groups of turbines with a capacity of 2 million kW per year will be installed.

Revenue from electric power generation

In Wang Jiazhu's article, the statement is made that while construction is going on, some 270 billion kWh of electrical energy will be produced with a gross output value of 18 billion yuan (at a per kilowatt value of 0.065 yuan) (Wang: 200). Assuming that the capacity utilization rate of 0.64 is correct (*N.B.* The rate assumed by Wang is higher than the 0.53 given by other authors), the sum total of the revenue generated would be 18 billion yuan over the seventeen years. One problem with this and most other economic assessments in China, however, is that it is assumed that a yuan earned today will have the same value as a yuan earned sometime in the future. This is in sharp contrast to the practice in Western countries, where the time preference of consumption or the discount rate is applied to all economic calculations. The discount rate is the societal value placed on receiving one dollar today rather than next year. Moreover, in Western economies when cost-benefit studies are undertaken, the streams of costs and benefits that may occur at different times in the future are reduced to their present value; otherwise they could not be accurately compared.

If we apply these principles to the revenues from the generating output of the TGP during its first seventeen years as described by Wang, and the 7 percent interest rate used in some of his calculations, we obtain the results given in table 6.

Using this standard approach to calculate the value of the electric power produced during the construction period, it is clear that the total gross revenues derived are not the 18 billion yuan presented by Wang but a more modest 6.5 billion yuan. We shall return to this point below, when we consider the overall costs and benefits of the TGP.

Table 6

Present Value of Electricity Produced at the TGP, Year 11 through Year 17

Year	Value of electricity produced (billion yuan)	Discount factor	Present value (billion yuan)
10	0	0	0
11	0.364	0.475	0.173
12	1.100	0.444	0.488
13	1.826	0.414	0.756
14	2.561	0.388	0.994
15	3.289	0.362	1.191
16	4.023	0.339	1.364
17	4.750	0.317	1.506
Total	17.913		6.472

Can the electrical energy produced by the TGP be used effectively?

In none of the papers translated is there any reference to the question of the effective utilization of the energy produced by the TGP. An electrical system must be designed in such a manner that it can cope with fluctuations of supply and demand. The power output of a single facility like the TGP will have a variable output of electrical power since water levels in the reservoir will fluctuate. There will also be daily and seasonal variations in demand by the users of electricity. The base load is the minimum below which demand never falls; peak demands occur for short periods only. If the output generated exceeds the demand at any time, revenue is lost because there will be no buyers (electricity, unlike other energy sources, cannot be easily stored). On the other hand, if demand exceeds supply, industrial output will be lost and the economy will be affected. Hence, a realistic assessment of the costs and benefits of the TGP cannot be made without some understanding of supply-and-demand fluctuations.

Clearly, if they are in phase there is no problem, but if they are not, information must be provided as to what will be done to remedy the situation.

In most countries, electricity supply depends not on a single facility but on several facilities using different sources of energy and linked by transmission lines, or interties, into an integrated system. The most economical way to operate such a system and ensure that supply meets demand is to use facilities with low operating costs, such as hydropower stations, running as nearly continuously as possible to meet at least the base load demand (Eden 1981: 185–89). When the base load demand is exceeded, the next block of demand is met by coal- or oil-fired plants, the operating costs of which are higher but whose flexibility allows them to be used for frequent shorter periods. Demands very close to the peak are met by gas turbines, which have high operating costs but even greater flexibility than traditional fossil fuel generators.

Unfortunately, in the literature reviewed and translated, information about the system design, the interties among different types of generating facilities, and the phasing of peak supply and demand is lacking, and hence it is impossible to arrive at a realistic assessment of the efficacy of the electric power component of the TGP. The possible losses involved in the above may reduce the benefits presented in the cost-benefit analysis.

Navigation

The principal concerns of navigation in the section of the Changjiang between the Three Gorges and Chongqing are that water levels remain as constant as possible; that the depth of the water be sufficient to cover shoals; that there should be freedom from siltation in water courses, particularly at port sites; that the rapid flow of the river be minimized because of haulage costs for vessels proceeding upstream; that narrow reaches of the gorge permitting one-way-only passage be eliminated; and that the locks on the dam permit the maximum tonnage of shipping to pass by without undue delays.

A large number of complex factors make it difficult to estimate the navigational benefits of the dam. Neither the CYJV study nor the various Chinese sources have dealt adequately with some important aspects. For example, the TGP is a multiple-purpose project involving power generation, flood control, and improved navigation. Generally speaking, for power generation and navigation, the higher the NPL the better. For flood control, on the other hand, it is necessary to draw down the water as far as possible prior to the rainy season to permit maximum reservoir storage capacity. The latter operating principle is clearly in opposition to the former. Moreover, because of the heavy sediment load of the dam, the impounded reservoir will be operated on the principle of storing the water when it is clear (i.e., at low-flow season) and discharging it when it is turbid (i.e., at the high-water season). This adds a further complicating factor. During the flood season when large quantities of water must be discharged through the dam sluices, it is not clear by how much water velocities in the gorge section will be reduced. Neither is it clear what impact the impounding of water during the dry season recharge of the reservoir will have on downstream flows between the dam and the Gezhouba reservoir. If the flow is too low, navigation may be impeded.

While many Chinese observers believe that the dam will bring substantial navigational benefits and convert the Changjiang into a "golden waterway" to the west (Wang, Qian et al.), others are less sanguine (Economic Construction Group, this volume). The latter group pointed out that some of the regulatory aspects of the reservoir described above will have an adverse effect on navigation. It is true that at the NPL 160-m level (the alternative recommended by the CYJV report), reservoir conditions will reach practically to Chongqing leaving only one one-way section as a potential bottleneck and eliminating all winching stations. This compares with thirty-two one-way sections and twelve winching stations under normal conditions (CYJV 1988: 1:12–6). However, according to the CYJV report (4:4–27), NPL 160 m will only occur for two months from November to December. From

June to the end of September, the reservoir will be maintained at the flood control level (FCL) of 140 m, and over 60 km of the river below Chongqing will remain in its normal state with at least five or six one-way sections remaining (CYJV 1988: 1:fig. 12.2). No consideration has been given to what effect these bottlenecks at low water will have on the larger tonnage of shipping predicted in the project.

There are also problems of siltation and gravel deposition and their effect on navigation that have not been addressed. These will be especially severe in the backwater reach downstream from Chongqing. The feasibility and costs of the massive dredging operations necessary has not been included in the estimates.

In addition, there is the question of the enhanced flow of traffic that will be generated as a result of the TGP. The Economic Construction Group, for example, remains skeptical that the one-way annual goal of 50 million tonnes of shipping projected by the Ministry of Communications can ever be reached. Even the usually sanguine CYJV study (1988: 1:12–7) suggested a low value of 27 million tonnes and a high value of only 38 million. Moreover, these values depend greatly on the size and number of vessels towed and the number of passenger vessels involved, since the latter will generally have priority, causing the freight vessels to be delayed. No thorough study has yet been undertaken of the tonnage that can be moved through the locks under various scenarios of vessel and tow size, the proportion of passenger vessels and the priorities to be assigned to different classes of vessel.

Finally, no adequate assessment has been made of the potential benefits to navigation from improvement and better traffic control in the gorge section without the TGP. Preliminary evidence found in the CYJV report (1988: 1:12–4) indicates that the present levels of 6 to 9 million tonnes could be raised to between 17 to 28 million tonnes depending on the size and mix of vessels. They also state: "the establishment of better traffic control procedures through one-way and winching sections or provision of additional aids to extend navigation hours could all serve to increase channel capacity beyond the limits defined above." If these improvements were implemented, the volume of traffic would equal or possibly exceed the low and high projections of the CYJV study with the dam. This would, of course, significantly alter the net benefits to be derived from improved navigation in the TGP. However, no estimates of this alternative have been given.

Overall costs and benefits of the TGP

Rogers (1986) pointed out in a symposium on the Three Gorges Project that "despite the intense study of the project by the Chinese government, there does not yet appear to have been any serious attempt to carry out an economic evaluation in a framework understandable to persons outside the Chinese government. . . . It would be helpful to know if the project is clearly a superior one from the economic point of view before spending the time and effort to make the detailed environmental studies which otherwise would not be required."

While we do not have the space nor the data to address this concern adequately, we shall attempt (following Rogers) to indicate some of the information and approaches required to carry out a viable cost-benefit study of the TGP.

The benefits of major water resource projects fall into two broad categories—current user values and intrinsic values—and each may be divided into direct and indirect. Current user values derive from the direct use of the water for transportation, fishing, recreation, or hydropower generation, or by withdrawing water for municipal, industrial, or irrigation purposes. Intrinsic values reflect the price people would be willing to pay to maintain a particular irreplaceable asset, such as the Three Gorges, or scenic site even if they did not use it. Stavins (1984) reviewed a number of water resource studies of user and intrinsic values and found that intrinsic values averaged 60 percent of user benefits, giving them a level of importance that must not be ignored in a valid TGP evaluation.

The distinction (seldom articulated in the TGP literature) must also be made between economic and financial appraisals of the project. The latter considers only those benefits and costs that directly affect the project; externalities, by definition, are not included in this type of analysis (Dixon and Easter 1986). Economic analysis, in contrast, is an accounting framework that takes into consideration the benefits and costs affecting society as a whole, and the negative and positive externalities, excluded in financial analysis, are now assessed.

As indicated above, a valid cost-benefit analysis of the TGP cannot add all the benefits and subtract all the costs as though they occurred at the same time to arrive at an estimate of the net benefit of the project. The main point to observe is that most of the costs occur at the beginning of the project, and the benefits begin later, when the project is completed (in the TGP case, seventeen years later). To make them comparable it is necessary to convert the stream of costs and benefits to one comparable year (i.e., their present value) at an appropriate social discount rate (SDR). Rather than using a specific SDR, agencies such as the World Bank normally use a measure known as the Internal Rate of Return (IRR), which is the discount rate such that the present worth of benefits is equal to the present worth of costs (Gittinger 1984). Since a precise level for the SDR is difficult to determine, one can compare the IRR with the notion of what the SDR should be. If the IRR exceeds this notional value, we can assume that the project is economically viable.

In a pilot cost-benefit study carried out by Rogers (1986), using very tentative data, the financial analysis gave an IRR of only 8 percent, whereas the notional SDR is around 10 percent. In the economic analysis, on the other hand, the IRR was 13 percent, an apparently viable project. However, if the calculations are carried out on the basis of coal-fired generation rather than hydropower, the economic IRR for the project falls to 8 percent, well below the notional SDR and therefore uneconomical.

Although the data on which Rogers' analysis is based were incomplete, his analysis suggests that doubts must be cast on the viability of the TGP from an economic standpoint alone. It is possible, however, that studies of the impact of the project on national and regional economies might reveal a greater ratio of benefits to costs. On the other hand, if the environmental impacts and intrinsic values of keeping the gorge area in its present state were included, the viability of the project could be further jeopardized.

The economic analysis of the TGP has been dominated by the benefits of power generation, transportation, and flood control. Since China has vast quantities of coal, and assuming that the environmental impacts of coal burning can be reduced, the planners of the project will have to produce more evidence that substantial benefits will accrue to other sectors of the economy before the TGP can be pronounced viable.

A full and valid CBA must also take into account the potential costs that have been identified in this introduction:

—Costs of raising the embankments below the dam in order to realize the flood control benefits of the TGP.

—Costs of dredging at Chongqing and downstream of the dam in the Gezhouba reservoir.

—Damage from floods in the Chongqing area due to elevated levels in the TGP reservoir.

—Possible damage caused by salt water intrusion in the Changjiang Delta region.

—Costs of giving adequate technical training to the population moved from the inundated areas so that they can benefit from the modernization of the infrastructure to be built in the reconstructed upland region.

—Possible loss of electrical power generation revenues from the failure to match supply and demand and inadequate interties among generating units using different sources of energy.

—Costs and benefits of alternatives to the TGP. For example, the CBA for a number of smaller dams instead of one massive dam should be evaluated.

Political and jurisdictional problems

David Lampton (1983) has pointed out that the top leaders of China do not unilaterally impose their will on the bureaucracy and territorial units when water resource decisions are made. Rather, there is an elaborate process of consultations and negotiations among bureaucratic and territorial units that can lead to interminably long delays. Moreover, as in most countries, there are institutional rivalries (Lieberthal and Oksenberg 1988: 290–338). According to Lampton, other ministries view planners in the Ministry of Water Conservancy and Electric Power and in the Changjiang River Basin Planning Office as flood control advocates. They are regarded with distrust by the Ministry of Communications and thermal electrical power interests. Water resource developments place great political strains on an entrenched territorial and bureaucratic system, as found in China, because they result in a redistribution of costs and benefits over which the units have no control (Lampton 1983). It is because they upset the status quo that water resource projects are usually regarded with great suspicion by these units. Such political strains may have led the Chinese government recently to consider the idea of creating a new province in the gorges area known as Sanxia Sheng (Three Gorges Province).

Conclusions

In our opinion, despite the major investment of time and effort in determining the feasibility of the TGP, serious conceptual and data shortcomings remain. From the cursory CBA presented above, it is by no means clear that the benefits of the TGP outweigh the costs. The large number of potential costs that have not been perceived or evaluated in the studies translated in this volume must be included in any adequate CBA for the project. These will undoubtedly make the scheme even less economically attractive.

As is often the case with large projects in other parts of the world, however, the symbolic benefits of the dam have a nonmonetary value that is perceived to exceed all monetary costs. Such may be the case in China, and the dam will be constructed regardless of monetary costs and benefits.

End Note

As we had anticipated, on March 20, 1992, the Three Gorges Project was approved in principle by the People's Congress, the highest legislative body in China. According to the news report available, the preferred project was NPL 175m. It is now up to the State Council to plan the construction schedule and it is expected that some work will start during the Ninth Five-Year Plan (1996–2000).

References

CAS. See Zhongguo kexueyuan.

Changjiang liuyu guihua bangongshi. 1979. *Changjiang shuili shilue* (Brief history of water conservancy in the Changjiang River Basin). Shuili dianli chubanshe.

Changjiang liuyu guihua bangongshi. 1988. "Sanxia gongcheng de fanghong zuoyong yu shengtai huanjing xiaoyi" (Flood prevention function of the Three Gorges Project and its impact on the ecological environment), in *Changjiang sanxia gongcheng dui shengtai yu huanjing yingxiang ji qi duice yanjiu lunwenji* (Collected papers on the ecological and environmental impacts of the Three Gorges Project). Shuili dianli chubanshe.

Chen Guojie. 1987. "Sanxia renkou yi chaozai, ziu di yimin shi zainan" (The TGP area is already overpopulated and local resettlement will be a disaster). *Shuitu baochi tongbao* (Soil and water conservation bulletin) 7, 5:42–46.

Chen Yueying. 1986. "The Three Gorges Project and Its position in China's Electricity Development" [quoted in Si], *Changjiang* 16, 5.

CPPCC. 1988. See Zhongguo renmin zhengzhi xieshang huiyi.

CYJV (CIPM Yangtze Joint Venture). 1988. *Three Gorges Project Feasibility Study*. Vols. 1–11.

Dai Qing. 1989. *Changjiang, Changjiang*. Guizhou renmin chubanshe.

Dixon, J. A., and K. W. Easter. 1986. "Economic Analysis at the Watershed Level," in *Watershed Resources Management—An Integrated Framework with Studies from Asia and the Pacific*, ed. K. W. Easter, J. A. Dixon, and M. M. Hufschmidt. *Studies in Water Policy and Management*, no. 10. Boulder: Westview Press.

Eden, R. J. 1981. *Energy Economics—Growth, Resources and Policies*. London: Cambridge University Press.

Fang Zongdai. 1989a. "Nisha chuli shi shuili gongcheng chengbai de guanjian, zhengque guji renlei huodong zengshazhi" (Sediment management is the key to the success or failure of water conservancy projects and there is a need to properly evaluate the impact of human activities on the increase of sediment production), in *Zailun sanxia gongcheng de hongguan juece* ed. Tian Fang and Lin Fatang, pp. 282–81.

Fang Zongdai. 1989b. "Sanxia shuiku dui Dongtinghu, Hankou de youhai fenxi he dui Chongqing zuigao hongshuiwei de weixie" (The negative impact of the TGP on Dongting Lake and the city of Hankou, and flood heights at the city of Chongqing), in *Zailun sanxia gongcheng de hongguan juece*, ed. Tian Fang and Lin Fatang, pp. 306–16.

Fearnside, P. M. 1988. "China's Three Gorges Dam: 'Fatal' Project or Step toward Modernization?" *World Development* 16, 5:615–30.

Gu Hengyue, Ai Nanshan, and Ma Hongliang. 1987. "Sanxia shuiku nisha laiyuan ji yanbian qushi" (Sediment sources and trends in the TGP reservoir area), in *Changjiang sanxia gongcheng dui shengtai yu huanjing yingxiang ji qi duice yanjiu lunwenji*, ed. Zhongguo kexueyuan, Sanxia gongcheng shengtai yu huanjing keyan xiangmu lingdao xiaozu, pp. 522–41.

He Naiwei. 1989. "Changjiang shangyou shuitu baochi he yingzao fanghulin tixi, guanxi dao zhongxiayou changzhi jiuan" (Downstream effects of soil and water conservation and protection forestry in the upper reaches of the Changjiang River), in *Zailun sanxia gongcheng de hongguan juece*, ed. Tian Fang and Lin Fatang, pp. 317–25.

Kennedy, J. F. 1987. "Sediment and Flood Control Aspect of the Three Gorges Project, China," *Keji daobao* (Science and technology review), special issue no. 1. (Original published in Chinese; English version provided by author).

Lampton, David. 1983. "Water Politics," *The China Business Review* (July–August).

Li Rui. 1985. *Lun sanxia gongcheng* (On the Three Gorges Project). Hunan kexue jishu chubanshe.

Lieberthal, Kenneth, and Michel Oksenberg. 1988. *Policy Making in China: Leaders, Structures, and Processes*. Princeton: Princeton University Press.

Lu Qinkan, 1989. "Fenxi Changjiang gezhong leixing hongshui jieguo shuoming, sanxia gongcheng de fanghong xiaoyi youxian" (Consideration of different flood types shows that the flood control benefits of the TGP are limited), in *Zailun sanxia gongcheng de hongguan juece*, ed. Tian Fang and Lin Fatang, pp. 299–305.

POCRB. See Changjiang liuyu guihua bangongshi.

Ren Rushu. 1987. "Guan yu sanxia gongcheng fanghong xiaoyi de wo jian" (Opinions on the benefit of flood prevention of the Three Gorges Project), *Keji daobao* (Science and technology review) 3.

Rogers, P. 1986. "Planning without Facts: A Framework for Economic Evaluation of the Three Gorges Project," *Keji daobao* (Science and technology review), special issue no. 1. (Originally published in Chinese; English version provided by author).

Shea, C. P. 1988. "Shifting to Renewable Energy," In *State of the World 1988, A Worldwatch Institute Report on Progress toward a Sustainable Society*, ed. L. R. Brown, pp. 62–82. New York: Norton.

Shi Deming. 1989. "Jiasu kongzhi Changjiang shangyou shuitu liushi shi xingjian sanxia gongcheng de qianti" (Speeding up the control of soil erosion in the upper reaches of the Changjiang River is a fundamental requirement for the construction of the TGP), in *Zailun sanxia gongcheng de hongguan juece*, ed. Tian Fang and Lin Fatang, pp. 326–29.

Shi Deming, Yang Yansheng, and Lu Xi. 1987. "Sanxia kuzhou diqu turang qinshi dui kuqu nisha laiyuan de yingxiang ji qi duice" (The impact of soil erosion on sediment sources in the TGP reservoir area and countermeasures), in *Changjiang sanxia gongcheng dui shengtai yu huanjing yingxiang ji qi duice yanjiu lunwenji*, ed. Zhongguo kexueyuan, Sanxia gongcheng shengtai yu huanjing keyan xiangmu lingdao xiaozu, pp. 498–521.

Si Zhizhong. 1986. Unpublished manuscript.

Stavins, R.. 1984. "The Toulumne River: Preservation or Development," Environmental Defense Fund, mimeo.

Sun Yueqi. 1989. "Changjiang liuyu guihua gongzuo de jiben yuanze" (Basic principles for the Changjiang River basin planning), in *Zailun sanxia gongcheng de hongguan juece*, ed. Tian Fang and Lin Fatang, pp. 124–34.

Sun Yueqi, Lu Qinkan, and Qiao Peixin, 1987. "Sanxia gongcheng hai duo li shao, bu rong qi shang ya xia, huo guo yang min" (The Three Gorges Project has more negative impacts than benefits), *Shuitu baochi tongbao* (Bulletin of soil and water conservation) 7, 4.

Tian Fang and Lin Fatang, eds. 1988. *Lun sanxia gongcheng de hongguan juece* (Discussions of macro decision making of the Three Gorges Project). Changsha: Hunan kexue jishu chubanshe.

———. 1989. *Zailun sanxia gongcheng de hongguan juece* (Further discussions of macro decision-making of the Three Gorges Project). Changsha: Hunan kexue jishu chubanshe.

Zeng Qinghua. 1989. "Sanmenxia shuiku jingyan shuoming: 'chu qing pai hun' de shuiku yunyong fangshi bu neng jieque sanxia kuwei de nisha yuji wenti" (Experience from the Sanmenxia Reservoir indicates that the operating mode of "storing the clear water and discharging the turbid" does not solve the problem of sedimentation at the backwater zone of the TGP reservoir), in *Zailun sanxia gongcheng de hongguan juece*, ed. Tian Fang and Lin Fatang, pp. 330–33.

Zhongguo kexueyuan. Sanxia gongcheng shengtai yu huanjing keyan xiangmu lingdao xiaozu (Chinese Academy of Sciences. Leading Group of the TGP Ecology and Environment Research Project). 1987. *Changjiang sanxia gongcheng dui shengtai yu huanjing yingxiang ji qi duice yanjiu lunwenji* (Collected papers on ecological and environmental impact of the Three Gorges Project and countermeasures). Beijing: Kexue chubanshe.

Zhongguo kexueyuan. Sanxia gongcheng shengtai yu huanjing keyan xiangmu lingdao xiaozu (Chinese Academy of Sciences, Leading Group of the TGP Ecology and Environment Research Project). 1988. *Changjiang sanxia gongcheng dui shengtai yu huanjing de yingxiang ji duice yanjiu* (Ecological and environmental impact of the Three Gorges Project and countermeasures). Beijing: Kexue chubanshe.

Zhongguo kexueyuan. Zhongguo ziran dili bianjizu (Chinese Academy of Sciences, Physical Geography of China Editorial Committee). 1981. *Zhongguo zirandili-dibiao shui* (Physical geography of China—surface water). Beijing: Kexue chubanshe.

Zhongguo renmin zhengzhi xieshang huiyi sichuan sheng weiyuanhui kaochazu (China People's Consultative Committee Sichuan Committee Investigation Group) 1986. "Changjiang sanxia gongcheng kuaishang manshang zhi zheng" (The controversy on the timing of the Three Gorges Project), *Shuitu baochi tongbao* (Bulletin of soil and water conservation) 6, 5.

1

A Review of the Work during the Early Stages of the Three Gorges Project

Hong Qingyu

1. Introduction

THE KEY to the Three Gorges Water Conservancy Project (TGP) is in the Xiling Gorge along the Three Gorges section on the main channel of the Changjiang (Yangtze) River. The proposed dam site is located at Sandouping, in the municipality of Yichang in Hubei province, 47 km upstream of the city of Yichang. The proposed dam will control a drainage area of about one million km^2, approximately 56 percent of the Changjiang River's total drainage area of 1.8 million km^2. At the dam site, the average annual flow is 450 billion m^3, approximately one-half of the river's total amount of 960 billion m^3.

The idea of constructing the TGP was first enunciated in Sun Yat-sen's plans for industrial and commercial development. During the 1930s, the Construction Committee of the Nationalist government investigated the Three Gorges and proposed the building of a low dam. In the 1940s, the Resources Committee of the Nationalist government signed an agreement with the U.S.

Hong Qingyu, ''Sanxia gongcheng qianqi gongzuo huigu,'' *Zhongguo keji shiliao* (China historical materials of science and technology) 8, 3 (1987): 3–10. The author is senior engineer, Planning Office of the Changjiang River Basin. Translated by Jane Tan.

Bureau of Reclamation to conduct further study of the proposal.

Since 1949, to control and develop the Changjiang River, large-scale surveying, planning, design, and scientific research in preparation for the project have been carried out. Planning for the comprehensive utilization of the Changjiang basin began in 1955. By 1958, the "Report on the Main Design Features of the TGP" had been completed. The report dealt with the overall planning of the Changjiang River and demonstrated that the project was of pivotal importance among the steps to be taken to control and develop the river. It showed the potentially significant impact of the project in preventing the threat of floods along the middle and lower reaches of the river, harnessing its water resources, and improving navigation in the river and its tributaries. In March 1958, at its meeting in Chengdu, the party's Central Committee discussed the TGP and the planning of the Changjiang River. In its resolution, the Central Committee pointed out that "from the points of view of long-term economic development, and of the technical capability of the country, the TGP is needed and can be constructed. . . . From now on, active preparation and sufficiently reliable policies must be adopted to facilitate all necessary work to construct the project." The resolution also stipulated that "all preliminary design work is to be submitted by 1962–63." The meeting was immediately followed by systematic surveys, design, and scientific research. The Central Committee's original thinking was to commence construction of the project in the early 1960s. But due to temporary economic difficulties and an unfavorable international situation, the project was later postponed. Nevertheless, following the spirit of Premier Zhou's instruction at the time to maintain determination and strengthen scientific investigation, all preparatory work was actively continued.

During the more than twenty years since, with the cooperation of the work units concerned, the Planning Office of the Changjiang River Basin (POCRB) has collected, organized, and analyzed a large amount of basic information. It has repeatedly carried out experiments on a number of major technical problems. The hundreds of work units associated with the project have written a large number of reconnaissance, survey, design, and research reports. They have organized numerous large-scale discussions and reviews by experts on major technical issues. In 1970, the Central Committee decided to commence construction of the Gezhouba (Gezhou Dam) "as a preparation for the actual battle at the Three Gorges." The construction of the Gezhouba allowed for the accumulation of rich experience needed for the building of the Three Gorges Dam. In 1983, after long years of investigation and comparative study of possible alternatives, and taking account of the socioeconomic conditions of the time, the POCRB completed a feasibility study report on a dam to store water to a normal depth of 150 m. In May 1983, the State Planning Commission examined the report, which was accepted in principle a year later by the State Council. The State Council instructed that a preliminary design be formulated for a dam to reach a height of 175 m and to store water to a normal depth of 150 m. However, differences of opinion still exist among the various groups involved. Taking account of the fact that the TGP is attracting world attention and is bound to have a major impact on China's four modernizations, the party's Central Committee and the State Council insisted that though much preparatory work had been done, valuable information obtained, and many experts consulted, there were still a number of problems and new proposals that should be evaluated from economic and technical points of view. The country's leaders recommended that based on a positive and prudent approach, and to achieve greater precision, reliability, and certainty, additional work on the feasibility of the project should be conducted. This work is currently being organized by the Ministry of Water Conservancy and Electric Power (MWCEP).

2. Major phases of strategic decisions affecting the TGP

2.1. Early study

The earliest reference to the plan to develop the Three Gorges may be found in the section on "Improvement of Navigable Rivers and Canals" in Sun Yat-sen's 1919 "Plan to Develop Industry." Sun proposed "to build a dam to store the water, so that ships can sail upstream against the flow of the river, and the river can be harnessed for electric power" at the Three Gorges. In 1924, in a speech on "People's Livelihood," he referred to the Three Gorges as "capable of providing more than 30 million horsepower of electric power." In the winter of 1932, under the sponsorship of the Nationalist Construction Committee, the National Defense Planning Commission organized a field reconnaissance of the hydropower potential of the upper reaches of the Changjiang River. Exploration and surveys were carried out in the Three Gorges area for approximately two months, and a report entitled "Reconnaissance Report on the Hydropower Generation in the Upper Reaches of the Changjiang River" was completed. The report proposed the construction of two low dams at Gezhouba and Huanglingmiao to produce 320,000 and 500,000 kW of hydropower, respectively. It is obvious that the improvement of the navigation channel along the Three Gorges and development of the water power resources of the Three Gorges have received people's attention for a long time.

In 1936, Bai Lang Du [transliteration; original name unknown—ed.], an Austrian adviser to the Yangtze River Water Resources Commission, also studied the problems of improving navigation and harnessing water power. His conclusion, however, was that "because of the depressed socioeconomic conditions, it would be especially difficult to carry out such a large project."

In April 1944, Pan Sui [transliteration], an American adviser to the Wartime Production Bureau of the Nationalist government, wrote a report dealing with the "utilization of American loans to construct a hydroelectric station and the repayment of such loans." He recommended the construction of a hydroelectric generating station at the Three Gorges to produce chemical fertilizers. Part of them would be used to repay the American loans, which would finance the purchase of necessary equipment.

In May 1944, John L. Savage, chief design engineer of the

U.S. Bureau of Reclamation, was invited by the Nationalist government's Resources Committee to visit China on his way to India. After being briefed on the conditions at the Three Gorges and on the report drafted by Pan Sui, Savage decided to make a personal visit to the Three Gorges. In September, on his return from the trip, at the hydropower project office at Longxi River, Changshou county, in the vicinity of Chongqing, he drafted the "Preliminary Report on the Yangtze River Three Gorges Plan," formally presenting a concrete plan to construct the Three Gorges dam. The report marked the beginning of a series of proposals for the development of the Three Gorges. It was followed by numerous negotiation sessions between the Resources Committee and the Bureau of Reclamation on technical assistance, leading to the signing of an agreement in 1946. In June, the Bureau of Reclamation officially began work on the design of the Three Gorges dam. The Chinese side also sent technical personnel to participate in the work. Altogether fifty-four people were sent, but only a small amount of exploration, survey, economic study, and design work was carried out. In May 1947, the Nationalist government decided to terminate the design work of the project. The predominant thinking during this phase was to develop the Three Gorges section of the Changjiang River to improve navigation and harness water power. There was not much actual work done, and that work did not have much impact on the decision to carry on with the project later on. Nevertheless, it served as the first revelation of the enormous potential benefits of the project to improve navigation and develop water power.

2.2. Formulation of strategic steps for flood control on the Changjiang River and the TGP

The Changjiang River is subject to floods from excessive rain. Potential flooding areas are very widespread, but the most frequent and serious floods occur in the plains of the middle and lower valley, where approximately 90 million *mu* of farmland, more than 75 million people, and the country's major industrial and cotton and grain production bases are located. Within this area, generally speaking, elevation is below the flood level of the Changjiang River. Protection from floods is provided by the building of embankments, which cannot completely remove the threat of flooding. Before Liberation, in certain plain and lake regions, floods were a regular occurrence. The Hubei folk ballad that mentions the frequent floods destroying crops in Mianyang in the central part of the province is a true depiction of the hazards of regular flooding in these regions. According to historical records, in the more than two thousand years from early Han to late Qing (between 185 B.C. and A.D. 1911) over two hundred floods occurred, averaging one flood per decade. As the population in these regions grew, and as the environment became further modified, the frequency of flooding increased. Between 1921 and 1949 there were seven major floods, with the destruction caused by the floods of 1931 and 1935 being especially serious. Affected were 50.9 and 22.64 million mu of farmland and 28.55 and 10.03 million people, respectively. In each flood, more than 100,000 people lost their lives. The 1949 flood was also a major one, and the population along the river had to be mobilized to struggle against the threat. There were significant losses.

The threat of flooding in the middle and lower valley is particularly severe along the section called Jingjiang (that is, the Changjiang main channel between Zhicheng and Chenglingji). The north bank of Jingjiang, known as Yunmengze in previous times, has now evolved into the Jianghan Plain. Dongting Lake on the south bank has shrunk in size. The embankments protecting the valley from flooding have had to be raised continuously. Although deemed to be somewhat unstable, the Jingjiang embankments along the north bank have reached an average height of 12 m and a maximum height of 16 m. In the event of a breach of the embankments, the death toll could reach 100,000, or even a million, and the security of Wuhan could also be threatened. This is the most pressing problem related to the prevention of floods on the Changjiang River. The problem attracted immediate attention after 1949 and led to the proposal of the Jingjiang Flood Diversion Project by the former Changjiang River Water Conservancy Commission in 1951. The proposal very quickly received the approval of the Central Committee and Chairman Mao, and construction began in 1952. The project was completed in seventy-five days after an intensive effort by 300,000 soldiers and civilians. It made an important contribution to the struggle against the major flood of 1954.

Based upon studies of the characteristics of the river, the Changjiang River Water Conservancy Commission put forward in the 1950s a three-phase improvement proposal which emphasized flood prevention. The first phase was to strengthen the embankments; the second, to reduce the scale of flooding by creating retention basins on the plains and in the lake region for flood water that could not be held back by the embankments; and the third, to construct dams in upstream mountain valleys for flood regulation and to develop water resources for multiple use. These three phases were not distinct but rather overlapped with each other.

The above proposal made the following points: (1) The primary and most urgent task in the improvement and development of the Changjiang River is the prevention of floods along the middle and lower reaches, especially the Jingjiang section. (2) Embankments are an effective means of flood prevention along the middle and lower reaches, and therefore as the first priority, they should be strengthened to give them better defensive capability. But because flood peaks on the Changjiang River are enormous, it is not realistic to combat floods by simply raising the height of embankments; furthermore, the higher the embankments, the greater the threat of floods. That the Jingjiang area is experiencing increasing threats of flooding is the result of poor flood prevention methods. Prevention of floods in the middle and lower reaches of the Changjiang River should be achieved through a combination of retention and discharge measures. (3) Controlling the lakes and depressions along the middle and lower reaches to store flood water is a relatively simple but effective measure. Every time flood water is stored, however, considerable damage is inflicted. The amount of damage will increase with the development of agriculture and industry in the area. It is therefore important to increase flood prevention capability upstream by building dams in mountain valleys. This will gradually replace flood retention in the plains or reduce the frequency of flood diversion.

Based on the three-phase plan, not only were the embankments strengthened and flood water storage on the plains provided, studies of the comprehensive utilization of the dams on the main channel and tributaries were also conducted. After comparative studies of various potential dam sites on the main channel and the tributaries, it was concluded that the most suitable site for the effective control of flooding in the middle and lower reaches was at the Three Gorges. The site was thought to possess the irreplaceable capability of preventing disastrous flooding in the Jingjiang area. In February 1953, after being briefed on the proposals to improve and develop the Changjiang River, Chairman Mao remarked: "If so much effort is needed to construct dams on the tributaries, and still it is not enough to achieve the objective of flood control, why don't we just concentrate on the Three Gorges and stop the floods there?"

During this period, flood prevention was clearly designated as the primary objective of the work to control and develop the Changjiang River. In studying flood prevention in the middle and lower reaches, especially along the Jingjiang section, the strategic position of the TGP was realized. The TGP is not only a hydropower generating station of enormous size, but also a multiple-use project that has a major role to play in flood prevention along the middle and lower reaches of the Changjiang River.

2.3. Formulation of the Changjiang River Basin plan and the 1958 Central Committee meeting in Chengdu

Flood prevention through the construction of dams in upstream mountain valleys involves the problem of multiple use of the water resources of the river and requires comprehensive planning for the river valley. In early 1953, the Changjiang River Commission established the Changjiang and Hanjiang River Planning Committee to begin plans for the two rivers. In 1954, the biggest flood of the last one hundred years occurred along the middle and lower reaches of the Changjiang River. Although the strengthening of the embankments and the completion of the flood diversion project along the Jingjiang following Liberation saved the Jingjiang embankments, the cities of Wuhan and Nanjing, and other key areas downstream, destruction was still very severe. Fifty million mu of farmland were inundated, and twenty million people were affected by the floods, with thirty thousand killed. This catastrophe revealed once again the seriousness of flooding along the middle and lower reaches and made people even more aware of the urgent need for establishing flood prevention as the primary objective of the plan to develop the Changjiang River. It also confirmed that the proposed three-phase development concept conformed to the real situation in the river, and that the final solution to flooding along the middle and lower reaches depended upon the construction of dams in mountain valleys upstream. The party's Central Committee immediately decided to commence planning, and Premier Zhou Enlai personally wrote to Premier Bulganin of the Soviet Union requesting that experts be sent to China to provide assistance. Soviet experts began to arrive in June 1955, and along with personnel from more than thirty Chinese work units, they began a program of systematic surveying, design, and study of the Three Gorges.

During the earlier period of planning, there were differences in opinion on locating the strategic key project. The Soviet experts were inclined to consider Maoer Gorge as the suitable location of the key project. Although a combination of projects at Maoer Gorge, Wentang Gorge on the Jialing River, and several other sites could solve the flooding problem along the middle and lower reaches, the inundated areas caused by the dams at these two gorges are much too large. We therefore favored the Three Gorges as the strategic key to the control of the Changjiang

River. At the end of 1955, after listening to the different viewpoints, Premier Zhou took the first step to define the Three Gorges as the principal part of the project for the whole Changjiang River Basin. The decision was based on the premise that the dam at Maoer Gorge would create an excessively large reservoir, while the Three Gorges "can regulate and store water for upstream, and provide compensation for downstream." Following this, more than a year was spent in further planning and study to demonstrate Chairman Mao's concept of "stopping the floods at the Three Gorges" and Premier Zhou's statement on the unique position of the Three Gorges in flood prevention along the middle and lower reaches.

In 1956, based on the preliminary results of the planning activities, Comrade Lin Yishan of the POCRB published a paper entitled "Discussion of Certain Problems Related to the Planning of the Changjiang River Basin" in the journal *Zhongguo shuili* (China water conservancy) (1956, nos. 5 and 6). He expounded the principles of multiple use in the development of the Changjiang River Basin, the major objectives, and the various developmental concepts and in particular discussed the role and effect of the TGP on the control and development of the Changjiang River. In September of the same year, Comrade Li Rui and others wrote in *Shuili fadian* (Hydroelectric generation) (1956, no. 9) a number of papers expressing disagreement with the view that attached such importance and urgency to the TGP in flood control on the middle and lower reaches of the Changjiang River, and doubts about the technical feasibility and applicability of the TGP. They recommended substituting Wuqiangxi along the Yuanshui River for the Three Gorges as the location of the key project.

In January 1958 in Nanning, during the Central Committee meeting, Chairman Mao, Premier Zhou, and other leading comrades listened to the different views of comrades Lin Yishan and Li Rui on the development of the Changjiang River. Chairman Mao stated that construction of the TGP should be based on the principle of "enthusiastic preparation and complete reliability."

He appointed Premier Zhou to take personal charge of the planning of the Changjiang River Basin and the TGP. Accordingly, between February 26 and March 5 of the same year, Premier Zhou led a group of more than a hundred people, including leading comrades in the central and local governments and Chinese and foreign experts, on an expedition to the Jingjiang embankments and the Three Gorges. He was briefed on the progress of the project and listened to views from different sides. At Chongqing, he presided over the meeting "to prepare enthusiastically for the construction of the Three Gorges Key Project," where more than twenty leading comrades and experts from different central ministries and commissions, provinces, and cities presented their views. Comrade Li Rui also participated in the expedition, and he spoke at the Chongqing meeting. He stated: "In terms of comprehensive benefits, the Three Gorges is ideal and very good. Although there are many technical difficulties, it must be said that it is feasible. In terms of meeting economic [efficiency] criteria, especially on a per unit input basis, the TGP also scores highly. There is no doubt that the TGP can solve the flooding problem on the Changjiang River and help in the development of water resources." He considered the controversy during the preceding two years as "mainly over the questions of the scheduling of the TGP, and the height of the proposed dam." In his summary at the meeting, Premier Zhou said that the central theme and purpose of the meeting was to study the question of "enthusiastically preparing for the construction of the TGP." He said: "Everyone unanimously confirmed the need and possibility of the TGP, its enormous political and economic significance, and its technical feasibility. It should be possible to be completed within not too long a period, between fifteen and twenty years." He added: "The debate over the past two years was also necessary. Without debate, how could we have obtained so much information to provide answers to every question that has been raised? . . . As long as controversy does not adversely affect our work, but contributes to it, it should be promoted and encouraged." The meeting brought to an end the more than two years of debate. It confirmed the TGP as the key project to the development of the Changjiang River, whose construction must be prepared with enthusiasm. For the TGP, this was undoubtedly the decisive strategic meeting. Based on the summary of that meeting, Premier Zhou presented the report "The Three Gorges Key Project and Planning in the Changjiang River Basin" at the Central Committee meeting held at Chengdu in March. On March 23, the meeting discussed Premier Zhou's report, and on March 25 it adopted the resolution "Opinion on the TGP and the Planning of the Changjiang River Basin." The resolution began by pointing out: "From the points of view of the long-term economic development and technical capability of the country, the TGP is needed and can be constructed. But the final decision to proceed with the project and the time to commence construction can only be determined after preparation work in the major areas is basically completed. It is estimated that the exploration and survey, design, and building will need between fifteen and twenty years. From now on, active preparation and sufficiently reliable policies must be adopted to facilitate all necessary work to construct the project." The resolution required that "attempts be made to submit the planning design of the TGP in 1959, and the preliminary design in 1962–63." It also stipulated that "the normal high-water level of the Three Gorges dam should be maintained at 200 m (above the Wusong bench mark [mean sea level]) and not any higher." It stated very clearly that the development of the Changjiang River Basin must be in accordance with the basic principle of "unified planning, overall development, appropriate division of labor, and proceeding in phases. At the same time, it should correctly resolve the following seven relationships: the long- and short-term prospects; the main channel and its tributaries; the upper and the middle and lower courses; the large, medium, and small scales; flood control, power generation, irrigation, and navigation; hydro and thermal power generation; and power generation and consumption (i.e., demand). These seven relationships should be properly balanced and, based on the actual conditions, determined whether they should be proceeded with what degree of emphasis, urgency, and order of priority." These principles put to rest such one-sided views as "tributary before main channel" and "small before large" expressed during the controversy.

2.4. Compilation of reports on the main features of the preliminary design of the TGP and extensive nationwide scientific cooperation

Based on Premier Zhou's summary at the 1958 Chongqing meeting and the resolution adopted at the Central Committee meeting at Chengdu, the POCRB began actively to carry out the preliminary design work.

The resolution at the Chengdu meeting confirmed, at the strategic level, the feasibility of the TGP. But the enormous size of the project meant that for certain major technical aspects, technology that surpassed the then world levels and established world precedents would have to be used. These aspects include: deep-water weir construction; damming under high flow conditions; super high voltage electrical transmission and transformation, super large turbo-generator and ancillary electrical equipment manufacturing; giant machinery for high-speed building and for elevating ships; automation in electrical systems and installations; new building materials; reservoir regulation; prevention of reservoir sedimentation, and issues related to economic planning. To solve these major scientific and technical problems, at the conclusion of the Chengdu meeting, Premier Zhou immediately established the TGP Scientific Research Group, made up of leading members of the State Science and Technology Commission (SSTC), the Academy of Sciences, and the MWCEP to provide leadership for the scientific research units across the country working on the project. In June 1958, the first Three Gorges Scientific Conference was convened in Wuhan to formulate the major research plans. More than two hundred work units and almost ten thousand scientific personnel took part in this nationwide cooperative effort. Scientific research during the two years between 1958 and 1960 generated a vast quantity of findings.

The first year alone saw the publication of more than seven hundred research reports and papers, which partially resolved a number of major technical issues of the project and provided scientific evidence for the preliminary design and technical design. Between the end of 1959 and the end of 1960, on the basis of the large volume of scientific findings, the POCRB completed the draft of the "Report on the Main Points of the Preliminary Design." In May 1959, the POCRB invited 188 people from sixty-six work units involved to discuss the report. Discussion was mainly concentrated on the two major issues related to the choice of dam site and normal water storage level. After ten days of discussion, it was unanimously decided that "the Sandouping site possesses unequivocal superiority." On the question of normal water storage level, it was thought that the key structure could be designed to a height of 200 m, but when the reservoir was put to use, the back-water level for the hundred-year flood at Chaotianmen should be kept below the 200-m mark, to prevent inundation of large areas of Chongqing.

In 1960, taking into account the country's economic conditions and the international situation, the party's Central Committee decided to slow down the planning of the TGP. In August that year, at the Central Committee meeting at Beidaihe, Premier Zhou personally chaired a discussion to rearrange the work of the TGP, instructing that "the determination shall not waver, scientific research shall be strengthened, national defence based on mobilization of the people shall be intensified." During the 1960s, in accordance with the spirit of the instructions from Chairman Mao and Premier Zhou, national defense based on the mobilization of the people and research on sedimentation was carried a step further. At the same time, based on domestic economic conditions at the time, a number of schemes to phase in the project were studied. This work continued up to the eve of the "Cultural Revolution." During this period (1958–1966), numerous technical issues of the TGP were settled, and the technical foundation of the project was laid.

2.5. Preparation for the real battle of the TGP: Construction of the Gezhouba

Between the last half of 1966 and 1969, the Cultural Revolution interrupted the long-term work to prepare for the TGP, bringing it to a halt. In October 1969, during his visit to Hubei, Chairman Mao was approached to revive the TGP by Comrade Zhang Tixue of the Provincial Revolutionary Committee. Chairman Mao pointed out that "At this time when the country is preparing for war, it is not wise to consider the project." Later, acting on suggestions from a number of comrades, Comrade Zhang proposed the construction of a project that is on a smaller scale than the Three Gorges—the Gezhouba, a component of the original TGP. Between the site of the Three Gorges dam and Yichang is a 40-km section of canyon. After the completion of the Three Gorges dam, navigation channels above the dam would be improved, but the 40-km section downstream would become the bottleneck of the navigation channel as it leaves Sichuan. Furthermore, because the outflow at the Three Gorges dam during the low-water period cannot be maintained at the regular level, it will have a significant daily variation and therefore will likely create navigational and berthing difficulties along this section and at Yichang. Through comparative studies, it was decided that these two problems can be solved by including the Gezhouba as a component of the TGP, to provide counter regulation. But the original schedule of construction called for the building of Gezhouba after or at the same time as the building of the Three Gorges dam. If Gezhouba were built before the Three Gorges dam, the degree of difficulty and construction cost would be increased, and there would be adverse effects during the construction of the Three Gorges dam. At the same time, there were still differences of opinion over the Gezhouba. After hearing the views of the different groups, Premier Zhou and the leaders concerned decided to go ahead with the Gezhouba project first because it was believed that it could serve as a preparation for the real battle at the Three Gorges dam. On December 25, 1970, when it approved the construction of Gezhouba, the Central Committee pointed out: "The construction of the Gezhouba Water Conservancy project is a preparation for the real battle to realize in a planned and step-by-step fashion Chairman Mao's lofty ideal of 'beyond the high dam emerges a vast and peaceful lake.'" On November 21, 1972, at the Gezhouba project briefing meeting, Premier Zhou again emphasized that "There was no real battle in the past; that is why we say we build the Gezhouba now as an experiment for the large Three Gorges dam."

Through the construction of the Gezhouba project, much valuable experience has been accumulated for the TGP. In the areas of sedimentation study and deep-water weir construction, for example, we have developed techniques that have given us more confidence and a better grasp of the problems of the TGP. An army of highly competent designers for large-scale water conservancy projects, construction workers, and scientific researchers have, at the same time, been trained. The Gezhouba has provided a reliable technical foundation for the TGP. Experience has shown that the Central Committee's attempt to use the Gezhouba as a preparation for the building of the Three Gorges dam was a wise and far-sighted decision.

2.6. Recommendations on the Three Gorges dam site

During the 1930s there was a suggestion that the dam site be located at Huanglingmiao. In the 1940s, John Savage selected six sites between Shipai at the exit of Xiling Gorge and Nanjin Pass (5–15 km from Yichang). These sites are underlain by limestone. After Liberation, the Huanglingmiao site was further investigated by the Upstream Bureau of the former Changjiang River Commission. By 1955, planning in the Changjiang River Basin and the design and study of the TGP was started. On the basis of previous studies, possible sites for the project were explored in two different sections of the river. The first was near the exit of the canyon section along the Nanjin Pass area, from Shipai upstream to Nanjin Pass downstream. This section of the river is 13 km long and is the section originally recommended by John Savage. The bedrock of the area is mainly limestone; thus it is referred to as the limestone dam area. The other is the Meirentuo section along the upper portion of Xiling Gorge, from Meirentuo to Nantuo, over a total river length of 25 km. The bedrock here is crystalline igneous rock. Thus, this is referred to as the crystalline or igneous rock dam area. The Huanglingmiao site proposed in the 1930s is located here. In both areas, a total of nearly 200,000 m of small-diameter rock cores were drilled. Over 150,000 m of this total was obtained from the igneous rock area, and about 76,000 m was from Sandouping. After comparisons were made, the Meirentuo area was chosen in 1958. This area was then divided into sections for further study. In the preliminary design proposed in 1959, the Sandouping section was chosen. This was followed by consideration of three dam sites along the section. By 1960, the upper site was finally chosen. This is the currently accepted site which passes through Zhongbao Island.

After 1960, a study of the air defense at Sandouping showed that the river was too wide to provide security at the site. Thus, investigation for another dam site was initiated. At the time, the suggestion was made to adopt the technique of massive directional detonation to build the dam, and to locate the power plants and navigational control facilities underground. For this reason, the Shipai site at Nanjin Pass was studied once again. After two years of research, it was found that construction at this site was more difficult, the time needed for completion would be longer, and it was not economically viable. Besides, the geological conditions were not good. The site was abandoned in 1963. Meanwhile, taking account of the need for air defense, the narrow stretch of the Meirentuo area was studied once again. Preliminary comparisons led to the choice of the Taipingxi section for more detailed study. Large-scale surveys produced the finding that the Taipingxi section possessed similar geological conditions to the Sandouping site chosen in the 1960s. Both are suitable for the construction of a high dam, but there are significant topographic differences. At the Taipingxi site, the valley is narrower. The dam requires less concrete to build, but the amount of excavation needed is larger. The Sandouping site, on the other hand, has a wider valley. It requires more concrete to build, but the amount of excavation is relatively small. In terms of the ease of construction, both have strengths and weaknesses and there is little difference in construction costs. Taipingxi is better for preventing air attacks, but not decisively when compared to Sandouping. Overall, both are good sites, but there was no unanimity as to which was the better. In 1974, after construction at the main site of the Gezhouba project was resumed, development of the TGP was once again being considered, and obviously siting of the dam was the primary concern. For this purpose, the POCRB compiled the "Report on Supplementary Studies on the Three Gorges Dam Sites" in March 1976. In November 1977, the MWCEP requested that the State Council call a meeting on the TGP, to facilitate the choice of a site and to promote a more focused research program. In February 1978, the MWCEP convened a "Preparatory Meeting for Site Selection." It was recommended that the State Council convene a site selection meeting, and that the POCRB compile a report on site selection according to the spirit of the meeting. In April 1979, the State Council called a meeting and received the reports on the progress of the TGP. It appointed Comrade Lin Yishan to preside over a "Site Selection Meeting." In May 1979, the site selection meeting at Wuhan was attended by more than two hundred representatives from fifty-five work units. At the meeting, it was agreed that both sites were suitable and a high dam could be built, but no unanimous decision was reached because opinions differed as to the pros and cons of each site. Some representatives expressed the view that it was premature to make the final decision and the Three Gorges Dam should not be built in the short term. After receiving the report of the meeting, the State Council instructed the former Ministry of Water Conservancy to convene a meeting to "report on the selection of dam sites." The meeting was arranged and was attended by some forty people, including group leaders at the Wuhan meeting, experts, and officials in leading positions. The meeting was again unable to reach a unanimous decision. After the meeting, the former Ministry of Water Conservancy presented a report to the State Council in November 1979. This report not only represented the different points of view, but also recommended that the Sandouping site should be selected for the preliminary design of the project. Thus, the long debate over the dam site was finally resolved, paving the way for more detailed survey and design and further study of the project.

2.7. Final decision on the site of the Three Gorges Dam

After the meeting to "report on the selection of sites," the former Ministry of Water Conservancy submitted to the State Council "Recommendations on the Three Gorges Water Conservancy Project," in which it recommended that "the TGP be considered as a major strategic construction under the country's four modernizations program, and that the project should be completed in the 1990s." In addition, "the early phases of the TGP . . . should be considered a key collaborative project between China and the United States."

Meanwhile, Li Rui and others of the former Ministry of Electric Power were opposed to the construction of the TGP in the short term. They have repeatedly expressed their views, either personally or as members of the former Ministry of Electric Power, to the leading members of the party's Central Committee. Their main arguments were: "The Three Gorges Reservoir has limited flood prevention capability, yet requires excessive amounts of investment"; the TGP is considered mainly on the basis of its power-generating capability; and "at the present time, the country's financial capabilities are obviously insufficient to shoulder the burden." To ensure consideration of these dissenting views, the State Council at its Standing Committee Meeting in August 1980 decided that "on the question of the construction of the TGP, the SSTC and the State Construction Commission should continue to organize evaluation meetings where views of experts in water conservancy, power generation, and other areas can be expressed." On the basis of this decision, the commissions proceeded to prepare a joint meeting and to request the POCRB to compile the necessary materials for discussion. At the time, the main argument against constructing the TGP in the short term was that a large number of people have to be resettled and the amount of investment is excessively high. Because of this, the POCRB investigated different schemes of phased development, in part based on research conducted in the 1960s. In January 1981, the POCRB proposed a scheme for phased development in which the first phase would have a water storage level of 128 m and a dam height of 145 m. At the end of 1981, the POCRB compiled and submitted a "Report on the Review of Evidence Supporting the TGP Proposals." The meeting to discuss this report has not yet been called.

In 1982, following instructions from the leadership, the POCRB began to study the feasibility of a storage level of 150 m. In March 1983, the "Report on the Feasibility Study of the Three Gorges Water Conservancy Project" was completed. The main argument in the report was that although the combined benefits of the 150-m scheme were less than those of a higher storage level scheme, these benefits were still very significant. "Undoubtedly, its potential economic return is superior," "considering its effect of tying up capital (in both cases) . . . it is still economically viable." "As to the major technical problems, they can be resolved after long years of study." "The issue of raising capital can also be settled," "thus, within the short term, there exist suitable conditions for constructing the TGP."

In May 1983, the State Planning Commission convened an evaluation meeting attended by 350 representatives from the concerned ministries and commissions under the State Council; the three provinces of Sichuan, Hubei, and Hunan; work units involved in scientific research, design, and construction; and from institutions of higher education. At the meeting, the great majority of the representatives agreed with the views expressed in the feasibility study report and advocated early commencement of the project. A minority disagreed, suggesting that conditions were still premature. Quite a few felt that the conditions for the TGP were superior, and the anticipated benefits enormous. They felt that while divergence of opinion existed, it was not related to major technical issues, and therefore a firm decision should be made immediately. The most controversial technical problem at the meeting was on reservoir sedimentation. Consequently a decision was made to study the problem through experimental research.

Following the review of the feasibility study report and based on the findings from the supplementary work, the MWCEP in February 1984 submitted a report to the State Council "recommending immediate commencement of the construction of the TGP." The report demonstrated the feasibility of the project, pointing out that the TGP enjoyed the following outstanding advantages: (1) the geographical location was a unique gift of nature; (2) the combined benefits were significant; and (3) development of energy resources was obviously necessary. The various major problems raised in the review report, such as reservoir sedimentation and navigation during construction, have, after a period of further study, been proven to be surmountable. The two points that the review report saw as needing further verification—the index for the inundation of material objects following the creation of the reservoir and the project cost—had also been positively verified. The Finance and Economy Group of the Central Committee examined the report submitted by the MWCEP. It agreed to prepare for construction in accordance with the scheme providing a water storage level of 150 m and a dam height of 175 m. In April 1984, the State Council formally accepted the view of the State Planning Commission expressed during its review of the feasibility study report. The State Council thus, in principle, approved the feasibility study report, instructing that the preliminary design should include a water storage level of 150 m and a dam height of 175 m. On the basis of this decision by the State Council, the POCRB immediately proceeded to carry out the scheduled work for the preliminary design phase of the project. In March 1985, it completed the report on the preliminary design. Thus, the preparatory work for the TGP has taken a giant step forward, providing the conditions for the commencement of construction.

2.8. *Supplementary review of the TGP*

Although technically feasible and economically beneficial, the 150-m scheme still does not provide sufficient combined benefits to satisfy its critics. At the review of the feasibility study report in 1984, a number of representatives suggested that the water storage level should be raised. During its preparation of the preliminary design, the POCRB was also approached by various groups to raise the level. Consequently, the SSTC and the State Planning Commission were asked to conduct a supplementary review of the issue. The POCRB was responsible for preparing a report for the supplementary review. At the same time, scientific research proceeded, especially with reference to sedimentation and resettlement. Review of water storage levels by the two commissions involved the study of different aspects: flood prevention, navigation, electric power system, sedimentation, inundation and resettlement, geology and earthquakes, and comprehensive assessment. Although the various specialist groups were yet to achieve a unanimity of views, the majority was inclined to consider the 150-m scheme as the most suitable.

During the review of the 150-m scheme, a number of dissenting views were also aired in some quarters, and doubts about the economic and technical feasibility of the TGP were expressed. This led to the issuance of a circular, jointly by the Central Committee and the State Council in May 1986, that the TGP would be reexamined. The circular pointed out: "During the recent period of over three decades, the country's ministries and departments concerned and scientific and technical personnel have carried out large numbers of surveys, scientific research, and design work to accumulate valuable information on the TGP. The State Council has also on numerous occasions organized expert discussions and has in principle approved the feasibility study of the TGP. But there are still a number of problems and new proposals related to the project that require further study into their economic and technical aspects . . . to enable the work on the Project to become more meticulous, accurate, and reliable." The MWCEP was instructed to organize experts for a further review. Accordingly, the ministry established a review leadership committee, under which ten special topical groups were organized. Four hundred experts divided into fourteen groups are now engaged in examining the feasibility of the project. They are to present their reports in 1987. We believe that through this supplementary review, a correct decision will finally be made.

2

The Features of the Three Gorges Reservoir

Tao Jingliang

THE KEY components of the Three Gorges Project are, as with many large-scale water conservancy projects in the world, main dam structure for water impoundment; an attached spillway structure, navigation structures, and hydropower plants located behind the main dam. However, the Three Gorges Reservoir formed after the construction of the dam possesses unique characteristics which differ significantly from other large reservoirs. These characteristics to a large extent determine the life of the reservoir, the degree to which the surrounding ecological environment is affected by the reservoir, its modes of operation, and the maximization of the project's benefits. Above all, these characteristics have a direct influence on an accurate appraisal of the advantages and disadvantages of the project. The following is a brief discussion of the salient features of the Three Gorges Reservoir.

1. Three Gorges Reservoir is a typical canyon-type reservoir

For normal pool levels of 150 m and 180 m, the length of the reservoir will be 560 km and 690 km respectively. In both cases,

Excerpts from Tao Jingliang, "Jianlun Sanxia Shuiku de Tedian," *Keji daobao* (Science and technology review) 2 (1987): 8–12. The author is deputy secretary-general, Construction Committee of the China Water Conservancy Society. Translated by W. X. Cheng.

the reservoir will be a narrow, ribbon-like structure, surrounded by low and high mountain gorges.

The Three Gorges of the Changjiang River refers to the section from the Sandouping dam site to Fengjie with alternating canyons underlain by carbonate rocks and wider river valleys of sandstones and shales. The imposing and precipitous cliffs towering above either side of the canyon section have an average height of 300 to 500 m, but the twelve peaks of the Wushan Mountains rise to as high as 700 to 1,100 m. The average width at the base of the canyon is 200 to 300 m. In the section of wider valleys, the height of the valley slopes is 100 to 300 m, the width of the valley bottom, 500 to 800 m. The Fengjie-Chongqing section of the Three Gorges Reservoir is located in the southeast margin of the Sichuan Basin, which is composed of low mountains and hills underlain by clastic rocks of Triassic and Jurassic age. The elevation of the mountains flanking the river declines gradually from between 700 and 800 m in Fengjie to 300 and 500 m near the Changshou-Chongqing area. Though the mountain slopes of the section are relatively gentle, the valley bottom remains quite narrow—only 600 to 1,000 m wide. Because of these limitations of topography, in the case of normal pool levels of 150 m or 180 m, the water surface width of the reservoir will be 800 m or 900 m to the east of Fengjie, and 1,200 m or 1,600 m to the west with a maximum width of 1,700 m or 2,000 m. Thus, there is no "bulging" part along the reservoir. In the 150- and 180-m schemes, the water surface areas will be 626 km^2 and 1,100 km^2, respectively. All the data above suggest that the 560–690-km-long reservoir has all the characteristics of a canyon-type reservoir. This is uncommon for large reservoirs.

In comparison, the Aswan Reservoir is located on the lower reaches of the Nile in southern Egypt. Its high dam is an 111-m-high earth core rockfill dam. Its main function is water storage, supplying the stored water to the lower streams as needed. The High Aswan Dam created a 500-km-long, 12-km-wide artificial lake, the second largest in the world. Some 350 km of the lake, known as Lake Nasser, lie in Egypt while 150 km, called Lake

Nuba, lie in the Sudan. The total water surface area of the Aswan Reservoir, a typical lake-type reservoir, is 6,500 km^2.

In the 150-m scheme, the length of the Three Gorges Reservoir is almost the same as that of the Aswan Reservoir. However, in terms of width and total water surface area, the latter is ten times as large as the former. It is clear, therefore, that the Three Gorges Reservoir and the Aswan Reservoir are two totally different types of reservoir.

2. The Three Gorges Reservoir is a typical reservoir that regulates seasonal river flow

Reservoirs can be classified according to the temporal scale at which they can regulate the river flow: seasonally, annually, or long-term. Usually, this regulatory function is assessed by considering the ratio of the storage capacity to the mean annual inflow.

If the elevation of the crest of the Three Gorges Dam is 175 to 185 m, the net height of the dam will be 165 to 175 m. Thus, for the normal pool levels of 150, 160, 170, and 180 m, the storage capacities will be 19.7, 26.2, 34.4, and 44.6 billion m^3, respectively. In terms of dam height and capacity, the Three Gorges Reservoir is certainly one of the largest in the world. However, it is far from big when compared with the mighty flow of the Changjiang River at the Sandouping dam site where the mean annual flow rate is 451 billion m^3 (based on the data supplied by the Yichang Hydrological Monitoring Station). This is 5.37 times that of the Nile and 6.55 times that of the Huanghe River (Yellow River).

When the normal pool levels of the Three Gorges Reservoir are 150 m, 160 m, 170 m, and 180 m, the ratio of the storage capacity to the mean annual inflow will be 1:22.8, 1:17.2, 1:13.1, and 1:10.1, respectively. Therefore, no matter which water level is adopted, the Three Gorges Reservoir can only regulate river flow in the Changjiang on a seasonal basis, with some differences in regulating ability at different water levels.

The storage capacity of the Aswan Reservoir is 164 billion m^3. The mean annual flow of the Nile at the dam site is 72 billion m^3. Thus, the capacity-inflow ratio is 2.28:1. That is, the Aswan Reservoir can hold all the water from the Nile over a period of two years. Therefore, the Aswan Reservoir is a typical reservoir that regulates long-term river flow.

From the two characteristics of the Three Gorges Reservoir described above and the comparison with the Aswan Reservoir, it is clear that these two reservoirs are completely different. Therefore, projecting problems that have occurred in the Aswan to the Three Gorges is scientifically unsound.

3. Three Gorges Reservoir is located in Central China, a superior geographical location favorable for realizing enormous benefits

The Changjiang River has a total length of 6,302 km and a drainage area of 1.8 million km^2. The main stream of the upper reaches (upstream from Yichang, Hubei) is 4,511 km long with a catchment area of 1 million km^2. The Yibin-Yichang section of the upper reaches is called "Chuanjiang" and is 1,030 km in length. The well-known Three Gorges section, 200 km long and stretching from Fengjie in Sichuan province to Baidicheng in Hubei, is located in the eastern part of Chuanjiang. The middle reaches of the Changjiang River, from Yichang to Hukou in Jiangxi province, is 938 km in length and has a catchment area of 680,000 km^2. The Zhijiang-Chenglingji section in the middle reaches is called "Jingjiang," and its channel is characterized by an extremely high sinuosity, especially in its lower section. Rushing out of the Three Gorges, the Changjiang River suddenly becomes wider with much slower flow velocity, causing significant alluviation on the river bed. Because of this and the many floods that occurred in the past, the Jingjiang section has developed natural levees with its channel elevated above the surrounding floodplain. It is only with the help of embankments on both sides of the river that the lives and property of thirteen million people have been protected. At present, by adopting emergency measures, the maximum regulated discharge at Zhijiang Hydrological Station section is 70,000 m^3/s. If the embankments are further heightened and reinforced, this rate can be increased to 80,000 m^3/s. Because of stability problems, the embankments cannot be raised indefinitely. If the discharge exceeds 80,000 m^3/s, the Jingjiang embankments will be in danger of collapse. So, it can be said without any exaggeration that "of the ten-thousand-*li* Changjiang, the Jingjiang section is the most dangerous part." At Dongting Lake, three distributaries of the Jingjiang enter, along with the four rivers of the Dongting drainage system. From here they flow into the main channel of the Changjiang River at Chenglingji. Thus, the Dongting Lake area is drained by a complex river system and is therefore another key flood control region along the Changjiang River. The section from Chenglingji to the estuary constitutes the lower reaches of the Changjiang River, a section that is 853 km long and has a catchment area of 120,000 km^2.

The Sandouping dam site of the Three Gorges Project is located at the middle-lower section of the picturesque Xiling Gorge, 37 km from the Nanjin pass, where the canyon section of the Changjiang River ends. This dam site controls the "throat" of the Changjiang River, which connects the narrow canyon section upstream to the wide river channels of the middle and lower reaches. Moreover, lying downstream of the region of intense rainstorms in eastern Sichuan, the dam will likely control more than 95 percent of the floodwaters in Jingjiang, and two-thirds of that at Hankou city. This favorable geographical location enables the Three Gorges Reservoir to develop fully its flood control potentials. In the event of a hundred-year flood (with a peak discharge of 83,700 m^3/s), it can protect the embankments at Jingjiang from collapse without diversion to the Jingjiang flood-diversion basin. Where a severe flood that has a frequency of once every thousand years or more occurs, the Three Gorges Reservoir, combined with the Jingjiang embankments and flood-diversion basin, can effectively protect the

Jingjiang section and the Dongting Lake area from flood disasters.

China is rich both in coal reserves and in hydropower resources, but the distribution of these resources is quite uneven. "Water in the south, coal in the north" is the generalization made by some. Of the proven coal reserves, 64 percent are in North China, 3.4 percent in the four provinces of Central China, and only 6.3 percent in the five provinces and one municipality of East China. Waterpower resources are also unevenly distributed throughout the country. The region to the west of the Beijing-Guangzhou railway line accommodates 90 percent of China's total waterpower resources. Of the country's total exploitable waterpower resources, 53 percent are held in the Changjiang River system, of which 88 percent are located upstream of Yichang. Yunnan, Guizhou, and Sichuan provinces and Xizang (Tibet) of southwestern China account for 67.8 percent of the national total; Central China, 11.9 percent; East China, 2.6 percent; and North China, a mere 1.2 percent. It is obvious that East China is very poor in both coal and waterpower resources.

The Changjiang River's near 100-m drop in the Sandouping-Chongqing section and its high flow rate will make the Three Gorges hydropower plant one of the largest in the world. With normal water levels of between 150 m and 180 m, the installed generator capacity could reach 13 to 18.72 million kW, with an operating capacity of more than 10 million kW. The annual working hours can amount to about 5,000, with an annual electricity generation of 67.7 to 89.1 billion kWh. The proposed Three Gorges hydropower plant is located in the heart of China, right at the "southern water and northern coal" and the "east-west electricity transfer" divides. The distances between the plant and such key cities as Beijing, Shanghai, and Guangzhou are about 1,000 km. It would be feasible to employ the 500 kv (ultra-high voltage) DC transmission line to supply electricity to these key cities, as well as regions in Central China, East China, and eastern Sichuan province. It would be particularly advantageous to employ a large hydropower plant to regulate supply in a large-scale power grid because this would allow the hydropower and the thermal power sources to supplement each other and the rational use of energy. If the Three Gorges hydropower plant is put into operation, 38–50 million tonnes of coal can be saved each year and the stress on railway transportation caused especially by north-south coal transfers can be greatly relieved.

The Changjiang River is the main artery of inland water transportation in China. From Yibin to Haikou, the 2,800-km water course is open to navigation all year round, whereas the 660 km section from Chongqing to Yichang has only been marginally improved despite major attempts to do so. Situated in an area of low hills, high mountains, and deep gorges, this latter section is characterized by a rugged topography, rapid currents, dangerous shoals, and narrow and meandering channels. At present, there exist 139 shoals, 25 towing stations for boats going upstream, 45 one-way channel sections, and 27 sections for daytime navigation only. The present condition of the Chuanjiang channel not only hampers the development of transportation here but also increases the transportation cost. Take the tugboat for example. In Chuanjiang, the towing efficiency is only 1 t/hp, the freight rate is 0.02–8 yuan/t-km, and accidents occur frequently. In the middle and lower reaches, the corresponding figures are 6–8 t/hp and 0.018 yuan/t-km; and 10–12 t/hp and 0.007 yuan/t-km, respectively. This means that the transportation cost in the upper reaches is about ten times that of the lower. Therefore, the above data suggest that to turn Chuanjiang into a real "golden waterway," there have to be fundamental improvements in the channel conditions. The Chuanjiang section is a channel that links Southwest China with Central and East China. So, the expansion of inland water transportation along Chuanjiang is of great significance to the economic development of Southwest China, a region which has rich natural resources and a large population.

The site of the Sandouping dam is located in the middle segment of the completed Gezhouba (Gezhou Dam) Reservoir. At present, the water depth exceeds 20 m here. Once completed, the Three Gorges Reservoir, when combined with the Gezhouba

Reservoir, will make stepped canalization possible in the shoal-stricken Chuanjiang section upstream of Yichang. In the 150-m, 160-m, 170-m, and 180-m schemes, the length of channel with adequate depth will be 585 km (to Changshou), 660 km (to Chongqing), 700 km (to Chajiao), and 730 km (to Lanjiatuo at Jiangjin), respectively. These will be fundamental improvements in the channel conditions of the Chuanjiang.

Once the Three Gorges Reservoir is completed, it will be possible to increase the flow rate in the lower reaches during the dry season. With increase in water depth, this will ensure unimpeded navigation from Hankou to Chongqing for 10,000-t vessels.

3

Comprehensive Assessment of the Ecological and Environmental Impact of the Three Gorges Project

Wang Chaojun

THE THREE GORGES Water Conservancy Project (TGP) is a colossal project that has caused public concern in China and attracted global attention in recent years. The Central Committee and the State Council have followed a positive and yet extremely prudent approach toward the final approval of the project. They have clearly stated that an updated report on the feasibility of the project should be prepared on the basis of detailed investigation and assessment. In the report, the assessment of ecological and environmental impacts is one of the most important components. Indeed, recent investigations conducted by the Bureau of Water Resources Protection, Changjiang River, and its associated research institute have been concerned with such major research topics as the selection of the normal water storage level for the reservoir and its impact on the ecological environment.

Wang Chaojun, "Changjiang sanxia gongcheng shengtai yu huanjing Yinxiang de zonghe lunzheng," in *Changjiang sanxia gongcheng shengtai yu huanjing yingxiang wenji* (Symposium on the ecological and environmental impact of the Three Gorges Project) (Shuilidianli chubanshe, 1988), pp. 3–17. The author is affiliated with the Institute of Water Resources Protection, Changjiang River. Translated by Jiaqing Gao and Ngan-chu Liu.

1. History of impact assessment

As early as the 1950s when the Planning Office of the Changjiang River Basin (POCRB) and other related institutions compiled a report on the general planning of the river basin and on the preliminary design of the TGP, several environmental factors relevant to the project were surveyed and investigated. These factors include the impact of backwaters, the effect of human activities on water discharge, reservoir slope stability, earthquakes, resettlement resulting from inundation, siltation, water quality, aquatic organisms, natural sources of virus [sic], and local infectious diseases. Some preliminary results were obtained and were included in the planning report. In 1979, the Institute of Water Resources Protection, Changjiang River (IWRPCR) was established as an associated institute of the Bureau of Water Resources Protection, Changjiang River (BWRPCR). The institute, in collaboration with more than forty institutions of higher education and research, organized research teams to investigate the environmental impact of the TGP. By 1980, a report on this subject was prepared for a normal pool level (NPL) of 200 m. In May 1983, a report on the feasibility of the NPL 150-m scheme was published. A chapter on "The Environmental Impact of Dam Construction at the Three Gorges" was contributed by the institute. During the period when the preliminary design of the TGP was conducted, two research reports, "The Impact of the Three Gorges on the Environment" and "Report on the Environmental Impact of the Three Gorges Water Conservancy Project," were completed. Other research reports on the impact of the project on water quality, climate, soils, vegetation, animals, and human health were also issued. In comparison with other major water resource projects around the world, all aspects of environmental impact of the TGP that should be considered have been evaluated. To ensure the quality of the evaluations, POCRB has secured the services of thirteen environmental scientists as consultants, who, together with other experts, were invited to attend and to present their opinions and recommendations in two symposia held in October 1984 and July 1985.

In May 1985 the State Science and Technology Commission (SSTC) convened a meeting in Chengdu to review the ecological and environmental impact of the TGP. During the meeting, members of BWRPCR presented a paper on "Evaluation of the Environmental Impact of the TGP at Different Water Storage Levels (150-m and 180-m schemes)." They have also participated in the review of ecological and environmental topics sponsored by the SSTC in June–July 1985.

In September 1986, a conference on the ecological and environmental evaluation of the TGP attended by invited experts was held in Nanjing, during which a document entitled "Outline and Schedule for the Evaluation of the Ecological and Environmental Issues of the TGP" was approved. Of the eighteen issues discussed, members of the IWRPCR have conducted research and presented assessments on the following five topics: (1) environmental conditions, assessment, and predictions in the areas affected by the TGP; (2) the impact of the TGP on the drainage of floodwaters in the plains and lake regions in the middle reaches of the Changjiang River; (3) the impact of the TGP on the ecological environment of the river estuary; (4) the capacity for resettlement in the reservoir area; and (5) the impact of the TGP on citrus production. The IWRPCR has also investigated the impact of the TGP on the white crane in Poyang Lake and the habitats of other rare migratory birds, the impact on the spread of schistosomiasis, and the methodology for the comprehensive assessment of the ecological and environmental impact of the TGP.

During two meetings on the ecological and environmental impact of the TGP sponsored by the SSTC in June 1985 and March 1987, it was generally agreed that the most important ecological and environmental problems affecting the feasibility of the project are inundation in the reservoir area and carrying capacity for resettlement. Subsequently, the BWRPCR responded by strengthening its research commitment on the issue of population carrying capacity of the environment. The problem for the reservoir area was analyzed and patterns predicted at the macro scale on the following assumptions: the entire project will be constructed in a single phase; water storage and resettlement will take place in stages; and a normal pool level of 175 m will be adopted. Research reports or interim reports on the above research were recently completed.

The TGP is a massive systems engineering project. Its ecological and environmental impact is also a very complex systems engineering problem. At present, evaluation is being conducted from the perspectives of ecology and environment, at several degrees of aggregation, and according to different scenarios, thereby providing scientific evidence for decision making at the macro level.

In evaluating the ecological and environmental impact of the TGP, the whole drainage basin, including the upper, middle, and lower reaches of the Changjiang River, is regarded as a megasystem. It comprises two subsystems, the biophysical and socioeconomic environments. After thorough survey and investigation, eight environmental components and thirty-two environmental factors were selected for study. By means of mathematical modeling, indices of special characteristics, simulation, and comparative analysis, the quantitative and qualitative changes in each factor caused by the construction of the project were studied. Based on the systems perspective and the need to evaluate environmental costs and benefits, evaluation of the total environmental impact of the project with and without remedial measures was attempted.

As the first stage of assessment has basically been done, evaluation work is divided at present into three different categories:

Category I. Issues that have been investigated and definite conclusions drawn. Only supplementary assessment is conducted. These issues include the impact on the local climate in the reservoir area, on water temperature and water quality, on aquatic organisms, on fisheries in the area, on terrestrial plants and animals, on soil conditions and agricultural ecology, on human health, on cultural and historic relics, and the environmental impact of the construction itself.

Category II. Complex and controversial issues such as the impact on the population carrying capacity of the environment, on the ground water in the lake and plains region in the middle reaches of the river, and on the ecological environment at the estuary. Research is being strengthened both in depth and in breadth.

Category III. Issues that are closely related to other problems, such as the triggering of earthquakes and the stability of the reservoir shores. These are being investigated by research teams where work is clearly divided but properly coordinated. For other issues such as the settlement of migrants, the impact on upstream flood damage, the impact on flood prevention in the middle-lower reaches, and the problem of resettlement due to reservoir inundation, investigation is being done by other groups.

2. Environmental and ecological impact of the Three Gorges Project

In varying degrees, the TGP will have an impact on the ecological environment of the reservoir area, the middle and lower reaches, and even the estuary of the Changjiang River. The effects are both harmful and beneficial, short-term and long-term, present and potential, and revocable and irrevocable. Part of the harmful impact can be mitigated or eliminated by certain remedial measures.

2.1. Ecological impact in the reservoir area

2.1.1. Impact on local climate. The reservoir will have obvious ameliorative effects on the climate in the surrounding areas, with slight increases in average temperature. At present, the mean January and July temperature is 3.4–7.2°C and 28–30°C respectively. The absolute minimum temperature is –4°C in the west and –8.9°C in the east (Zigui County). In comparison, it has been estimated that after the construction of the reservoir, the winter and spring temperatures will increase by about 0.8°C, while in the summer, it will be lowered by about 1°C. The absolute minimum temperature in the reservoir area will likely be 3°C higher. The long-term average number of foggy days exceeds thirty in the area west of Wanxian County, and in Chongqing it is sixty-nine. After construction, conditions affecting fog formation and thus its frequency of occurrence will not change to any significant degree. At present, wind is slight, with an annual mean speed of 1–1.5 m/s. This may increase by 15 to 40 percent after construction. Relative humidity will also increase by 2 to 8 percent over the current 65 to 85 percent. The annual precipitation will hardly be affected, but its spatial distribution in local areas around the reservoir will be somewhat altered.

The rise in winter temperature creates favorable growing conditions for citrus during the cold months. The increase in instantaneous wind velocity will adversely affect river navigation.

2.1.2. Impact on water temperature and water quality. Normally, temperature stratification occurs in the reservoir water in early April. It is estimated that the temperature of the water released between February and May will decrease by 2.8–4.4°C (4.1°C in April). The period during which the released water reaches a temperature of 18°C will lag by about twenty days, which may adversely affect the reproduction of the four well-known domestic fish species.

Overall, water quality may be expected to improve because of the filtering effect of the reservoir. However, as flow velocity is reduced the dispersion of pollutants is less effective. This, in conjunction with the concomitant industrialization and urbanization in the reservoir area, will lead to worsening of pollution along the shores. It is estimated that the maximum concentration of pollutants will increase by 20 to 30 percent. Therefore, control and management of waste disposal from towns, factories, and mines should be strengthened.

2.1.3. Impact on aquatic ecology, aquatic organisms, and fisheries. After construction of the dam, the structure and function of the aquatic ecosystem will be considerably altered. The species and quantity of zooplankton and phytoplankton will increase. Species composition will likely shift from riverine to lacustrine. The density of primary aquatic organisms may significantly increase. Because of the increase in water surface area, development of fisheries in certain tributaries and estuaries in the reservoir area will become favorable if effective measures of fish culture are adopted. Between Chongqing and Zigui there are eight spawning sites for domestic fish. They will be completely or partially submerged after the reservoir is filled. Domestic fish will then have to migrate upstream to spawn in the upper sections of the reservoir or tributaries. Between Yichang and Chenglingji there are twelve spawning sites, accounting for 45 percent of the total along the main channel of the Changjiang River. After construction, the spawning environment will be adversely affected. At the Gezhouba (Gezhou Dam), natural reproduction of the Chinese sturgeon, which is a migratory fish species, was observed downstream of the dam. Artificial reproduction of the species has also shown some initial success. To prevent the extinction of the sturgeon, artificial reproduction should be expanded, and fishing should be properly controlled. Variations in water discharge and processes of scour and fill downstream of the dam may affect in varying degrees the habitat of the white flag dolphin.

2.1.4. Impact on terrestrial plants. In the proposed reservoir area, large forest areas do not exist below 200 m. Thus, if the 175-m scheme is adopted, natural vegetation communities will not be severely affected. However, orange groves (*Citrus reticula*) will be the major cash crop affected by inundation. Some 73,900 mu of orange groves will be inundated, most of which are located in western Hubei, and a disproportionately large area of 28,200 mu is in Zigui County. The impact on plants of economic value will be accentuated as such human activities as resettlement and construction begin. Preventive measures should be implemented to minimize such detrimental effects.

Most rare plant species in the reservoir area are found at higher altitudes. The extinction of lowland species can be avoided by transplantation or artificial cultivation.

2.1.5. Impact on the terrestrial animals. At present, the quantity and number of species of wild animals in the reservoir area are decreasing. After construction, the surface water area will be enlarged, thus creating a favorable environment for such rare animals as the mandarin duck and the otter. This will also lead to an increase in the number of amphibious animals. Birds and wild animals will migrate to higher areas. However, the density of small animals, especially the rat, will soar in farming areas. This will affect people's lives and the production of crops.

2.1.6. Impact on soil conditions and agricultural ecology. The reservoir will inundate an enormous land area leading to increased pressure on arable land. Soil erosion will be exacerbated by the resettlement if the environment is not protected. The original agricultural ecosystem must be restructured and ultimately replaced by a new one.

2.1.7. Impact on the geological environment. The reservoir area has a tectonic structure characterized by weak earthquakes. Assuming the worst scenario of a continuous fault zone and maximum energy release, an induced earthquake along the Xiannushan-Jiuwanxi fault located close to the reservoir site may reach a magnitude of 5.5 on the Richter scale, which translates into an earthquake severity of 6.0 at the Sandouping dam site.

The Three Gorges is an area of frequent rockslides and landslides. Recently, landslides that severely affected navigation and caused great losses of life and property occurred at Jibazi in Yunyang County and in Xintan. According to recent systematic studies of the environmental geology of the area, it was established that the weakening of the Gorge walls caused by impoundment will not be significant enough to threaten navigation and the structures related to the project. However, in selecting the sites for resettlement, the hazards of slumping and landslides must be fully taken into consideration. Monitoring of the areas where potential landslides may affect navigation should be strengthened and warning systems should be installed.

The rocks in the reservoir area are impermeable and no problem of water seepage is expected.

2.1.8. Impact on human health. According to recent investigations, no oncomelania or schistosomiasis is currently found in the reservoir area. However, suitable habitats for oncomelania and favorable conditions for the dispersion of schistosomiasis are found in some local areas. Thus, environmental changes in the area should be investigated, and the conditions for transmission should be controlled and monitored. After construction, the hazards of flooding and waterlogging in the middle and lower reaches of the Changjiang River will be mitigated. This would be helpful in preventing and controlling the dispersion of schistosomiasis in the Jianghan and Dongtinghu plains.

Once the reservoir is impounded, breeding of *Anopheles* mosquitoes along the mainstream is unlikely, but breeding is favorable at the confluence of tributaries and local depressions. Appropriate measures are required to prevent the outbreak of malaria.

To prevent and control other diseases, especially haemorrhagic fever in the reservoir area, attention should be paid to mice known to be carrying the virus. After impoundment, they will migrate to the higher areas. This will pose a serious threat to human life, and preventive measures should be strengthened. Moreover, attention should be paid to other diseases with natural dissemination sources such as paragonimiasis and schistosomiasis to prevent their transmission.

2.1.9. Impact on cultural relics and historic sites. The Three Gorges is a key scenic area in China. After impoundment, the following cultural relics and natural features will be submerged: Zhangfei Temple in Yunyang County, the "stone fish" in Fuling, the karst landscape in the Xiling Gorge and on the banks of the Daning River (such as "ox's liver and horse's lung" and "military books and treasured swords"), the carved writings on the walls of the Qutang Gorge, the Dual Temple in Jiefang, Quyuan Temple in Zigui, the Qiujiang Pavilion in Badong, and other well-known historic buildings and rock formations along the river.

With a higher water level and a broader water surface, the attractiveness of the gorge will be diminished in sections where the relative relief is limited. In most sections where the relative relief is high, the aesthetic appeal of the gorges will remain intact.

2.1.10. Environmental impact of construction. The TGP involves massive construction work, a high degree of mechanization, and concentration of manpower. There are a great number of potential problems, such as shore pollution, air pollution, noise, and changes in the local landscape. Diseases such as malaria and hepatitis may occur. The sanitary condition of the construction site and the health of the construction workers should be monitored and protected.

2.2. Impact on ground water in the plains and lake region in the middle reaches of the Changjiang River

The section of the Changjiang River between Zhicheng and Chenglingji is called "Jingjiang." To the north of the river lies the Jianghan Plain, which is characterized by fertile soil and abundant rainfall. It is one of the most important grain production bases in China. The four-lake region (Changhu Lake, Sanhu Lake, Bailu Lake, and Honghu Lake) is a low-lying hinterland of the Jianghan Plain, criss-crossed by rivers and canals and known as the "land of plenty" in Hubei province.

At present, the bed of Jingjiang is 1–2 m higher than its surroundings. During floods, the water level is 6–13 m above the adjacent land. This imposes a flooding and waterlogging hazard in the area which is aggravated by the coincidence of the flood and rainy season. Traditionally, the lakes in the Jianghan Plain have been important regulators of runoff, flooding, and waterlogging. However, the lake areas have been dramatically reduced due to years of alluviation and overreclamation. Their regulatory capacity has been depleted while the frequency of floods, droughts, and waterlogging has been greatly increased. The expansion of cultivated fields into low-lying lake areas has led to the deterioration of drainage conditions and the rise of the water table, causing waterlogging in fields of the lowland lake areas. To mitigate these problems, pumping stations have been constructed between Chenglingji and Xintankou.

The four-lake region is a broad alluvial plain, the level of which is coincident with the elevation of the first fluvial terrace. It is underlain by Quaternary deposits. The region is surrounded by water bodies on three sides and by hilly areas on the remaining side. Its elevation decreases from northwest to southeast, ranging from 24–35 m at its highest point to 21 m at its lowest. At present, the elevation of areas along the Changjiang River is higher than that of the hinterland for a number of reasons, including previous breaching of embankments along the Changjiang and Hanjiang rivers and the problem of siltation. The hydrogeological characteristics of the region are closely related to the regional geology and geomorphology. According to the analysis of drilling logs, ground water in most parts of the region has two layers: an upper layer of phreatic water (or shallow ground water) and a lower layer of artesian water. They are separated by an impermeable layer which is 5–15 m thick. The phreatic zone, which occurs entirely in Holocene deposits that show complex changes in properties due to the shifting of river channels and lake locations, has a thickness that varies from 3 to 8 m. Its depth below the surface varies with both topographic conditions and season. Actual data indicate that its depth is not only controlled by rainfall but also related to irrigation methods and the capacity of the drainage system. During the flood season when rainfall is abundant, the depth of ground water in the alluvial plain along the Changjiang River is 1–3 m, and in the lowland lake region, 0.2–0.4 m. In winter, its depth within the lakeshore region is around 0.5 m. The relationships between the ground water level at the Qigenwa area, Jianli County, and both the water level at the Changjiang River and the total ten-day rainfall have been examined. With a small time lag, ground water level rises in response to rainfall, but it is not significantly affected by the fluctuation of the river water level.

The artesian zone occurs in Quaternary deposits of alluvial sand and gravel which are overlain by argillaceous material located 10–30 m below surface. However, because of the range of flow conditions under which these argillaceous deposits are formed, their thickness varies greatly, with a maximum of about

20 m. In comparison, the artesian zone is 90–110 m in thickness. The artesian zone is recharged either laterally or vertically by other aquifers. During the flood season, when the water level at the Changjiang River is higher than that of the artesian zone, recharge by river water occurs. In the dry season, due to the lower river level, reverse flow occurs. However, the aquifer only dips gently at a gradient of about 0.7 percent away from the river. Thus, the hydraulic gradient is small, flow velocity is low, and the level of river water has little influence on artesian water level. This is demonstrated by data from six observation wells located near the Guanyin Temple at the Jingjiang embankments where the river water level is considerably higher than the adjacent land. Continuous monitoring since the early 1970s shows that the influence of river water level decreases away from the river. Beyond 2,438 m, changes in the artesian water level are not in phase with the river water level.

In addition, through the analysis of the coefficient of soil permeability (*K*), it was found that the surface material in the embankment area which has a clay loam or sandy loam texture has an average *K* value of 1×10^{-5} to 4×10^{-5} cm/s. The subsurface material is fine sand whose *K* value has an average of 3×10^{-4} to 4×10^{-4} cm/s. This shows that the velocity of ground water flow is very slow. If the dry season lasts for at least 150 days, the distance affected coincides fairly well with the above observed range. In general, the water level of the Changjiang River has little impact on the groundwater level in the four-lake region.

After the construction of the TGP, the water in the four-lake region will be drained by the Xintankou pumping station. In years with moderate precipitation, the Changjiang River water level will rise by about 0.75–1.0 m during January–April. Elevation of the area near Xintankou ranges from 22 to 24 m. The river water level is about 0.5–5.4 m higher than that within the four-lake region. But the water level just beyond the pumping station is higher than the water level behind the embankment. Therefore, the construction will not substantially affect the drainage of flood water and ground water. In May, the actual and projected mean water levels of the river are higher than those within the four-lake region both before and after construction. In October, because the reservoir is allowed to fill, the amount of regulated flow is less than the preconstruction discharge. This will require advancing the date for opening the sluice gate of the station and is conducive to the lowering of the water table. After construction, because the amount of displaced water will be reduced due to the rise in the water level of the river, the operating schedule of the Xintankou pumping station and small ports in Honghu County will have to be modified. In January and February when the water level of the river is the lowest, the capacity of the Xintankou pumping station should be utilized to the fullest so that the water levels of the major drainage canals in the four-lake region and of Honghu Lake can be lowered, thus expediting the rapid lowering of the water table in the cultivated areas. It is also possible to speed up the process by using the recently constructed Xintankou station and other electric pumping stations along the river. The additional cost will be minimal.

The above discussion shows that the construction of the TGP will not raise the water table in the four-lake region, and gleization and paludification will not be accentuated. On the other hand, the construction will alleviate the burden of local flood control and offer large quantities of inexpensive electricity. It will then be possible to increase drainage facilities, improve soil conditions, modify inappropriate irrigation and drainage methods, and reconvert some paddy fields back into lakes. In so doing, the hazards of flooding and waterlogging in the region can be mitigated.

2.3. Impact on the ecology and environment at the estuary

2.3.1. Impact on water discharge at the estuary. After construction of the TGP, the annual water discharge into the East China Sea in years of both moderate and low flow will remain unchanged. However, its seasonal distribution will be altered. For instance, at the Datong hydrological station, located near the estuary and unaltered by tidal water, discharge will slightly decrease during October–November but will increase during January-May due to the regulation of the reservoir. No change is anticipated during the other months of the year.

If the 175-m scheme is adopted, the discharge in October will be reduced by 20 percent in the years with normal rainfall and by 17 percent in the years with heavy precipitation. However, runoff at the Datong station will still be greater than 33,000 to 40,000 m^3/s. This discharge is still within the observed range for the same month prior to construction. In dry years, the reservoir stores water in October and November, thus reducing runoff at the station by 5,450 m^3/s in October and 3,000 m^3/s in November. This is equivalent to a discharge reduction of 32.4 percent and 18 percent in October and November, respectively. Statistical analysis of the daily discharge records from 1972 to 1984 suggests that, during October–November, the frequency of occurrence of a discharge less than 16,000 m^3/s at the station will increase from 2.12 to 3 days per month. Therefore, even during dry years, flow reduction in October–November due to the presence of the reservoir will have little impact on water discharge in the estuary. However, the above discussion does not take into account the additional discharge resulting from the improved scheduling of the regulated flow and the compensatory effect of flow from tributaries and lakes downstream of the dam. Because the station is 640 km away from the estuary, which is one-third of the total distance from the Three Gorges to the estuary, the additional discharge from a drainage area of approximately 100,000 km^2 would certainly increase the discharge at the estuary.

During January–May, the estimated regulated flow will increase by 1,000–2,000 m^3/s, and thus the dry season discharge at the estuary will be increased. By referring to data of the January–April period recorded from 1972 to 1984, it can be shown that discharge below 10,000 m^3/s at the station during October–November will be reduced from 10.58 to 6.23 days per month.

Such a reduction is significant from the perspective of water quality because dilution will be effectively increased, the concentration of chlorides will be reduced, and the frequency of occurrence of pollution will be diminished.

2.3.2. Impact on sediment at the estuary. The impact of the Three Gorges reservoir on sediment transport and on water discharge in the lower reaches of the river is different. During the initial period after construction, inflow and outflow of sediment at the reservoir will not be balanced. About 70 percent of the sediment will be trapped and 30 percent will be exported. Because a discharge of less than 60,000 m^3/s will flow through directly, and even for higher flows, only peak flow will be slightly reduced, changes in the sediment carrying capacity of the flow will be minimal. Thus, water issuing from the dam with a low sediment concentration will scour the downstream river channel. Eventually, farther downstream, sediment concentration will rise to a level compatible with the capacity of the flow. A similar tendency has been observed at a number of reservoirs both at home and abroad. For instance, at the Sanmenxia Reservoir, of the 6.76 billion tonnes of sediment reaching the reservoir from September 1960 to October 1964, 2.17 billion tonnes were trapped. But 2.31 billion tonnes were scoured from the channels below the dam, yielding a total of 4.48 billion tonnes at the estuary, which is approximately 66 percent of the sediment reaching the reservoir. Under natural conditions, the sediment delivery ratio at the channels of the lower Huanghe River is 0.75. Thus, without the dam, the total sediment discharged into the Bohai Sea would have been 5.07 billion tonnes during those four years. In comparison, sediment load at the estuary was still 88.4 percent of that discharged without the reservoir. Thus, as long as the amount of water discharged from the reservoir is not lowered, and peak flows are not excessively reduced, there will be minimal changes in the sediment load of the flow. In addition, a large portion of the middle and lower reaches of the Changjiang River is located in alluvial plains where sediment available for scouring is abundant. As a result, water issuing from the Three

Gorges reservoir during the initial period of operation will scour the 1,800-km-long channel downstream. Thus, the total sediment load discharged into the East China Sea will not be significantly reduced. Over a long period, say fifty years, it is expected that the sediment delivery ratio at the reservoir will reach 0.8. After eighty years, it will be higher than 0.9, by which time most sediment will be exported, a situation that is virtually the same as before construction. In general then, the Three Gorges Reservoir has a limited regulating effect, most of which is seasonal. It will have minimal effect on water discharge and sediment load in the middle and lower reaches of the Changjiang River and its estuary. Therefore, no major impact on the ecology and environment of the estuary is expected.

2.3.3. Impact on the incursion of sea water. The degree to which the incursion of sea water would occur depends on the discharge of the Changjiang River. The higher the discharge, the smaller the area affected by sea water and vice versa. Data on mean monthly salinity recorded from a ship-based marine hydrological station located 67 km downstream of Wusongkou show a high negative correlation with the mean monthly discharge measured at the Datong hydrological station. The period of high salinity tends to occur during December–April, with the peak in February. June–October is a low salinity period, with the minimum in July. After construction, because of the increase in regulated flow, the level of salinity at the marine observation station in February will decrease by 1.3 percent even during a dry year. In fact, the salinity level at the estuary of the river throughout January–May will be improved.

In Shanghai, the water for industrial, agricultural, and domestic uses is mainly derived from the Huangpu River, along which eight waterworks are located. Those waterworks affected by salt water incursion are Wusong, Zhabei, and Yangpu. They are respectively 1.7 km, 7.1 km, and 21.7 km away from Wusongkou. At present, the international and domestic water quality standard is 250 ppm of chloride for potable water. Usually, a figure of 100 ppm indicates that the water quality is affected by salt water incursion. When the chloride level exceeds 500 ppm, the water becomes a serious hazard for irrigation and domestic uses. Data collected from 1972 to 1985 show that when the discharge at the Datong hydrological station exceeds 20,000 and 18,000 m^3/s, the probability for the chloride level at the Wusong waterworks (which is closest to the estuary) to be >100 ppm and >250 ppm respectively is extremely low. Moreover, the mean discharge at the Datong observation station in October over the last forty years is 35,700 m^3/s. After construction, the regulated flow will be reduced by about 8,400 m^3/s during years with normal precipitation. Even so, the discharge at the station will exceed 20,000 m^3/s for 89 percent of the time and the water quality standard at the Wusong waterworks can be met. However, in dry years when the discharge at the station in October is <18,000 m^3/s, the salinity level in the Wusong waterworks will exceed 250 ppm. Statistical data show that during the October–November period when impoundment at the reservoir occurs, the frequency of occurrence of discharge >18,000 m^3/s will be reduced by ten days per year. The exceedence probability will be reduced from 95 percent to 80 percent. Thus, the storage of water in the reservoir during dry years will have considerable impact on the water quality at Wusongkou. However, the probability for this to occur is not high, and such impact can be reduced in the future by adjusting the scheduling of water regulation at the reservoir.

After construction, the average regulated flow in January-April will be increased by 1,000–2,000 m^3/s. During the same period, the mean monthly discharge at the Datong station is 8,900–20,700 m^3/s in dry years. Thus, the effect of the increased discharge on reducing salinity is obvious. Apart from reducing the peak chloride concentration, the frequency of occurrence of chloride concentration at the Wusong waterworks >100 ppm and >250 ppm could also be reduced by approximately ten and fifteen days respectively.

2.3.4. Impact on the nutrient availability at the estuary and on fish resources of the adjacent seas. Due to its enormous drainage basin and extensive river network, the Changjiang River transports a tremendous amount of nutrients which are inputs into the water body for various physical, chemical, and biological processes that occur naturally, and for various human activities. A relatively large portion of the nutrients transported in tributaries and through lakes eventually enter the East China Sea. They become the major food sources for marine organisms there.

The effect of the reservoir on nutrients is to intercept them, resulting in local deposition. However, the effectiveness of interception is closely related to the water transit rate within the reservoir. Because the Three Gorges Reservoir is designed to regulate seasonal flow and thus will maintain a high transit rate, the impact on interception of nutrients especially during the flood season is insignificant. Moreover, only 20–30 percent of the nitrogen and phosphorous loadings are derived from upstream of the dam, and the bulk of nutrients entering the estuary originates from an extensive area in the middle and lower reaches of the river basin. Therefore, the construction of the dam will not have an obvious negative impact on the availability of nutrients at the estuary.

Because of the large volume of available nutrients, a favorable marine climate, and geographic location, the estuary and the adjacent seas of the Changjiang River are important fishing grounds in China. After construction of the reservoir, the regulated flow will be increased during January–May. Considering the life cycle of the fish species in the estuary, most will migrate to the outer seas during the winter months of January and February. At that time, marine organisms are inactive due to the low temperature, and their demand for nutrients is low. Thus, changes in water discharge and nutrient availability have little effect on the fishing grounds during this season. However, during March–May when the temperature ameliorates, the fish begin to migrate to the estuary and the adjacent seas to spawn and feed. An increase in discharge brings more nutrients and is favorable for the increased productivity of fish species in the estuary and the adjacent seas, as well as to the species that migrate upstream to spawn. Thus, on the whole, the impact of the increased regulated flow on fisheries at the estuary and adjacent seas is mainly positive. In October–November, the regulated flow is reduced, and so are the nutrients transported. But by this time, winter is approaching and the plankton population is dwindling. Many fish species are on the verge of migrating to the outer seas. The demand for nutrients is also reduced. Thus, the negative impact of the decreased discharge on fisheries in the estuary and the adjacent seas is minimal during this season.

In addition, the decreased discharge in October may cause the shifting of the convergence zone of the fresh and sea water by several kilometers, thus resulting in the displacement of the fishing ground as well. Such displacement will be greater in dry years but would not be sufficient to cause the complete disappearance of the fishing ground. The fish habitat in coastal areas is more strongly affected by the distribution and shifting of the injected bodies of fresh water. However, changes in the location of the injected fresh water bodies is mainly influenced by black tides, the warm current from Taiwan, and northern cold currents. The effect of water discharge from the Changjiang River is subsidiary.

From the above discussion, it can be concluded that the TGP will have a considerable impact on the ecology and the environment of the river estuary, but major changes are not anticipated.

3. Population carrying capacity of the environment for the resettlement area of Three Gorges Reservoir

After the Three Gorges Reservoir is impounded, vast areas of land will be flooded and the large number of people displaced will have to be resettled locally but at higher elevations. The population carrying capacity of the reservoir area plays an important role in determining whether the local economy will continue to develop, whether degradation of the ecological environment can be avoided, and whether the people displaced can be satisfactorily resettled. This is a crucial factor in determining the success of the project.

The population carrying capacity of a region refers to the size of population that can be maintained at specific production levels, living standards, and environmental quality. It is related to the natural environment, in particular the region's natural resources and socioeconomic bases.

The population carrying capacity of the environment is determined by its level of productivity. It reflects the extent and efficiency of the exploitation of natural resources. With increasing productivity, the human ability to transform and harness nature will improve and the population carrying capacity of the environment will gradually expand. However, such expansion has limits because it is constrained by conditions of natural resources, the total environment, and the ecosystem.

3.1. Present population carrying capacity of the Three Gorges area

The proposed Three Gorges Reservoir is located at the junction of Sichuan and Hubei provinces, bisecting the southern foothills of the Daba Mountains and the hilly area in western Hubei. The topography of the reservoir area is complex. It is characterized by high mountains, steep slopes, incised river valleys, and high relative relief. Mountains and hills account for over 90 percent of the area, while the remainder are river valleys and floodplains. With poor agricultural resources and weak industrial foundation, the area has long been the target for state support. The poor economic foundation and limited investment have led to slow economic development and widespread poverty. Most old cities and towns of the area have an irregular street pattern. The streets are narrow and the population densities high. At present, the reservoir area remains economically backward, with poor environmental conditions and a relatively small population carrying capacity.

In the nineteen counties and municipalities along the shores of the proposed Three Gorges Reservoir, there is a total population of 13.585 million, an area of 54,200 km^2, and a population density of 251 persons per km^2. This density is more than double that of the national average.

The total cultivated land in the reservoir area is 14.187 million mu, or 1.04 mu per person, which is about two-thirds of the national average. Of the existing cultivated land, 60 percent is slopeland under dry farming, one-quarter to one-half of which has a slope gradient exceeding 25°, and "hanging slopeland" in which "harvest is left at the mercy of God." In Wanxian County, for instance, slopeland steeper than 25° accounts for 25 percent of the total cultivated land area. It is 40 percent in Kaixian County, 47.9 percent in Yunyang, and 53 percent in Wushan. If the slopeland is converted into forest land, per capita acreage of cultivated land will be substantially reduced.

The level of grain output in the reservoir area is low, with yields averaging merely 208 kg per mu, which is 15 kg less than the national average and less than half of that in the lower and middle reaches of the Changjiang River.

The total grain output in the reservoir area was 5.18 billion kg. The per capita output was 327.5 kg in 1984, which is less than that in Hubei province by more than 135 kg, in Sichuan by more than 75 kg, and even the national average by over 50 kg. The area has been barely self-sufficient in recent years.

Present forest cover in the reservoir area is low. Apart from several counties in western Hubei which have forest cover exceeding 25 percent, Wanxian County in eastern Sichuan and the counties and municipalities located along the reservoir shores in Fuling Prefecture have only 5–10 percent cover. Soil erosion is serious, and the ecological environment is undergoing degradation that is self-perpetuating.

The gross value of industrial and agricultural output in the reservoir area was 7.4 billion yuan in 1984, of which 3.68 billion was industrial output and 3.72 billion agricultural output. The per capita output value of industry and agriculture was 545 yuan, which was 354 yuan less than the national average of 899 yuan. The per capita output value of industry was 271 yuan, which was 323 yuan less than the national average of 594 yuan.

The low level of productivity directly limits the population carrying capacity of the environment. The excess supply of urban and rural labor is related primarily to the production conditions of society.

Based on these conditions, it is clear that the major reasons for the relatively small population carrying capacity of the reservoir area at present are the low level of productivity, its backward economy, and the incompatibility of economic development and population increase.

3.2. Impact of reservoir impoundment on the carrying capacity of the environment for resettlement

According to the 175-m scheme, the TGP will inundate areas affecting nineteen counties and municipalities. The upper tail of backup waters would reach Mudong in Baxian County, 38 km downstream from Chaotian Gate in Chongqing. The reservoir will have a surface area of 1,028 km^2, a length of 600 km, and an inundated area of 604 km^2.

Of a population of 714,500 to be displaced, 54 percent are urban residents and 46 percent rural. Kaixian County will have to resettle 92,000 residents, the largest number compared to other localities. Four counties and municipalities will have to resettle 60,000–80,000 people each; five counties to resettle 40,000–60,000 people each; two counties to resettle 10,000–30,000 people each; and seven counties to resettle less than 10,000 people each. Of the population displaced by the reservoir, about 114,200 are already residing in the twenty-year flood zone. This is because the flood prevention standards in the riverine cities and towns were unduly low.

Though the absolute number of people displaced is considerable, they are distributed along the banks of the 600-km-long reservoir, averaging 595 persons per km of shoreline. In total, they account for 5.3 percent of the total population of the nineteen counties and municipalities. Moreover, the impact of resettlement on population density is small, with an expected increase of only two persons per km^2.

After impoundment, an area of 348,300 mu of cultivated land will be inundated, accounting for 2.46 percent of the total cultivated land of the nineteen counties and municipalities and including 1.4 percent of the total paddy field area. The per capita cultivated land area will be reduced from 1.04 to 0.97 mu. An area of 73,900 mu of citrus groves, or 6.7 percent of their total, will be inundated. Zigui County will be hardest hit with 56.5 percent of its existing citrus groves area inundated. In Badong, it is 24.5 percent.

Most cultivated land affected is largely in river valleys with favorable soil and water conditions, and high multiple-crop indices and yield. According to one estimate, the total loss of grain output caused by inundation will be about 150 million kg, or 2.8 percent of the total grain output of all counties affected. This is equivalent to a loss of 44 million yuan, or 2.1 percent of the agricultural output value in the reservoir area.

Local facilities inundated by the reservoir will also include 620 industrial and mining enterprises, with a total fixed asset of 804 million yuan. In addition, 957 km of roads, hydropower and thermal power installations with a capacity of 67,000 kW, and 1,133 km of high voltage power transmission lines will be flooded.

The major enterprises affected include those involved in food processing, textiles, machine-building, chemical, coal, electric power, and building material. Because their foundations are weak, production facilities poor, and technology primitive, the level of production of these industries is very low. If the monetary compensation resulting from inundation is to be invested in technical transformation and technological renewal, the productivity of the industries will expand considerably and the level of output will improve substantially.

In general, the construction of the reservoir will inundate part of the existing resources, causing considerable damage to production facilities and the life-support system. Because of the large number of people displaced and the weak economic foundation of the area, resettlement to higher elevations will exert greater pressure on the land. If effective measures are not adopted, the ecological environment will be degraded, affecting, to a certain degree, the population carrying capacity of the environment. While the difficulty of resettlement is immense, the population to be displaced is scattered, and the proportion of the population to be displaced in the nineteen counties and municipalities is not excessive. These conditions are favorable for the organization of resettlement.

3.3. Approaches and measures of expanding the population carrying capacity of the reservoir area

Of the total population to be displaced, 54 percent are residents of cities and market towns. This nonagricultural population will be resettled en masse, according to the existing cities and work units. There will be no employment problem as they will retain their existing jobs, and the supply of commodity grain will be maintained. After several years of surveying, sites for relocation of most of the cities and market towns have been selected and relocation planning is underway. In the course of site selection, it was found that the available sites are higher than 182 m, and work involving large amount of land leveling and massive construction is required. These are not simple tasks given the constraints posed by existing topographic conditions. Nevertheless, suitable sites are available and construction in the resettlement area is planned according to the existing national standards. Housing and urban facilities will be substantially improved, residential population density will be reduced from 17,000 to 12,500 persons per km^2, and the built area will be increased by 50 percent. Also, the environmental conditions will be greatly improved.

Of the total population to be displaced, 46 percent is engaged in agriculture, primarily in farming. After inundation and resettlement, they may either continue as farmers or new jobs will have to be arranged. This is the crux of the problem when investigating the population carrying capacity of the Three Gorges area. If the residents in the resettlement area could be assured of an income equivalent to or higher than the original level and of a prospect of continuous improvement, then the problem of resettlement capacity can be resolved.

The following section is a discussion of the preliminary analysis regarding the conditions for expanding the population carrying capacity of the reservoir area.

3.3.1. Abundant resources and high potential for economic development in the reservoir area. The Three Gorges reservoir area is one of the regions in China with relatively abundant natural resources. As far as its natural conditions are concerned, the reservoir area is a typical humid intermediate subtropical zone, where the maximum amount of sunlight, heat, and rainfall received occur in the same season. Besides, topographic variations are conspicuous. All these provide favorable conditions for the growth of various living organisms. According to statistics from local sources, there are 10 million mu of unused hill-land and grass slopes, which can serve as land reserves for the development of macro agriculture (*da nongye*). Granted that large portions of the relatively level and good-quality land will be used for establishing new cities, towns, and villages, part of the low-lying unused hill-land and grass slopes do have great potential to be exploited as farmland and economic forest land. However, land use must be determined according to local needs and conditions, and soil and water conservation and farmland capital construction work must be strengthened. This is to prevent the degradation of vegetation and the ecological environment, and the intensification of soil erosion. Also, there are 9.6 million mu of unforested land which can be used to develop forests and to make up for part of the losses caused by inundation.

The reservoir area is rich in various mineral resources worthy of exploitation. Preliminary surveys show that they include gold, copper, iron, rocksalt, natural gas, barium, mercury, phosphorous, aluminum, and so on. There are also large amounts of building material resources such as marble and limestone. The reserve of rocksalt is the largest, now estimated at 150 billion tonnes, which can serve as an important material for salt-related industries. Coal is widely distributed over the area. At present, there are small coal mines in almost every county, with an average annual output of 200,000 to 500,000 tonnes each. The reserves of iron and phosphorous are abundant but are extracted only on a small scale. The reserves of natural gas are about 250 billion m^3, while the proven reserves of other resources such as barium are 46 million tonnes, one of the largest ore bodies in China. It is both necessary and practicable to establish in a planned manner a group of industrial, mining, and township enterprises after the completion of dam construction. In particular, labor-intensive manufacturing industries should be established to absorb part of the resettled workers and to develop the local economy. Certainly the development of industrial and mining enterprises should not cause pollution.

The Three Gorges area, which is endowed with numerous charming mountains and rivers, scenic spots, and historic relics, is a world-renowned tourist area. While part of the area has been developed, many tourism resources have yet to be exploited because of limitations of transportation development. After the construction of the Three Gorges Dam, improvement in transportation and the emergence of new cities and towns will undoubtedly create favorable conditions for exploiting the area's tourism resources. Hence, the development of a comprehensive tertiary industry emphasizing tourism will be fostered. As tourism develops, construction, building material, and transportation industries will grow correspondingly, providing more employment opportunities for the resettled workers. Moreover, there are large amounts of hydropower and aquatic resources to be exploited.

Rational exploitation and utilization of different resources in the area can transform resource potential into economic gains. The existing population carrying capacity of the environment can thus be expanded, and the ecological environment will improve.

3.3.2. Develop the area's strong points, adjust the agricultural and industrial structures, and expand employment chan-

nels. Rural migrants in the reservoir area should continue as farmers. Based on the resource, topographic, and soil conditions of the area, we should adjust the geographical distribution of crops and the industrial structure; develop macro agriculture; diversify the rural economy by developing farming, forestry, and animal husbandry according to local conditions; develop a superior economy; and build up a base for producing agricultural commodities. According to surveys conducted by Chinese and foreign experts, the region bordering the 2,000-km-long river channel from Yibin in Sichuan to Jiujiang in Jiangxi has a mild climate, with abundant rainfall, a long frost-free period, and fertile soil. It is an ideal region for growing fruits such as *Citrus reticula*. In the Three Gorges area, which is located within this region, citrus groves are cultivated at present on 1.1 million mu of land. The total output amounted to 156,000 tonnes or 150 million yuan in 1985. In fact, growing citrus has become one of the economic pillars of the reservoir area. Moreover, the potential for further development remains high. Based on the development plans of various counties, the acreage of citrus will increase by 720,000 mu and 1.22 million mu in the years 1990 and 2000, respectively. The annual total output will reach 310,000 tonnes and 1.1 million tonnes, or 300 million yuan and 1 billion yuan, respectively. Auxiliary enterprises involved in preservation, processing, and transportation will be established correspondingly. Not only will they raise the income of the in-migrants, but they will also create jobs for a large number of in-migrants. This is the most promising measure in developing the local economy. Fuling in Sichuan has enjoyed a long history of producing preserved vegetables, the operation of which has been integrated from the stage of raw material production to research to processing and marketing. To enhance the advantages of producing preserved vegetables, we should strive to forge the links of commodity circulation and open up the international market. The production of tung oil in this area has already accounted for one-quarter of the country's total and ranks first in export. However, there is still room for further expansion in export of processed tung oil. Precious medicinal herbs produced in the area are another item that enjoys good prospects for expansion.

Relatively large areas of unused hill-land and grass slopes are found in each county of the reservoir area. Raising dairy cattle and other herbivores through the growing of pasture in this type of land and land on the edge of the reservoir can yield high economic returns. Certainly, effective measures should be adopted and farmland capital construction and scientific farming methods strengthened to increase yield. In short, measures adopted should be suited to the local conditions, bringing into full play the potential of the local resources. People displaced should be resettled through different methods. Lateral economic integration within the province and outside, in China and abroad, should be strengthened to activate economic development.

3.3.3. Implement special policies beneficial to the economic development of the reservoir area. The TGP, which involves massive resettlement, is a project that relates to many areas of policy making. The project has an important bearing not only on the welfare of the people to be displaced, but also on the area's future economic development and long-term stability. It is therefore necessary to adopt a resource-exploitative resettlement policy, which integrates the construction, resource exploitation, and economic development of the reservoir area with the task of resettlement, and which constitutes an organic component of the entire construction project. In so doing, greater economic, ecological, and environmental benefits will be obtained, and certain special policies beneficial to the economic development of the reservoir area can be implemented. For instance, the area inundated can be linked to the benefits of power generation of the TGP. That is, for a specified period, part of the revenue from power generation can be transferred to the reservoir area as construction funds (e.g., at 0.2 cent per kWh, a total of 60 million yuan can be transferred annually). Also, the inundation of large amounts of land will result in a decline in grain output and a change in industrial structure which will cause conversion of part of the agricultural population into nonagricultural population.

This will require a corresponding increase in the supply of commodity grain. It would be desirable to provide tax exemption or to apply other preferential policies to agricultural, industrial, and commercial sectors to support the local economy. The cultural standards of the residents resettled should be raised through investment in education.

3.3.4. Resettlement should be compatible with the area's economic development and ecological environment. With a relatively backward economy and fragile ecological environment, the reservoir area is facing a resettlement problem that is a very complex social and ecological issue. Therefore, the construction design of the project should also incorporate a comprehensive plan for resettlement. The issues of resettlement, economic development, land management, and ecological and environmental protection should all be taken into consideration in a unified and rational manner. In developing urban and township enterprises, pollution control should be carried out to maintain ecological balance. In the upper reaches of the Changjiang River and in the reservoir area, soil conservation programs should be properly implemented to prevent soil erosion. It is important to develop a new and improved system of human ecology, an agreeable, ever-improving environment which can meet the productive, domestic, and ecological requirement of the residents.

In general, resettlement can follow the principle of relocating existing administrative units (i.e., districts, townships) upslope. This does not mean, however, that all the people displaced will be "packed" into the narrow strips of land fringing the banks of the reservoir. To prevent overcrowding in the fringe area of the reservoir, and the consequence of a drastic decrease in per capita acreage of cultivated land, resettlement at some localities may be carried out according to regions. That is, within the same county, resettlement areas may expand longitudinally where appropriate, resettling people in planned districts or townships where larger areas of usable land and richer natural resources are found and where the population density is low. In so doing, the necessary resources available for a complete reconstruction of a production and socioeconomic system with high potential for future development will increase. Therefore, resettlement will not adversely affect the current production, living standards, and ecological environment in the resettlement areas.

In short, the reservoir area possesses rich natural resources and high potential for further development. As long as there are appropriate and effective policies as well as sound resettlement plans for resource-exploitative migration, the carrying capacity of the resettlement areas may be considered acceptable from a macro viewpoint. Specific plans in different counties and some typical migration experiments have proven that this is viable.

4. Functions of the project and its major social and environmental effects

A project's ecological and environmental effects should be taken into consideration when carrying out a comprehensive environmental impact assessment of the project. This was recommended at an international conference on large dams and has been adopted by some developing countries.

The TGP serves multiple purposes, including flood prevention, power generation, navigation, aquaculture, and tourism. It will bring about enormous economic, social, and environmental benefits.

4.1. Effective reduction of flood hazard in the middle-lower Changjiang Plain; Improvements in production and living conditions

The middle and lower reaches of the Changjiang River are frequently ravaged by severe floods. Historical records show that there were more than 200 flood disasters of various severity during the two millennia from the Han to the Qing dynasty. This is equivalent to one flood disaster every ten years. In this century alone, severe floods occurred in 1931, 1935, 1949, and 1954. Floods destroyed farmland and houses, causing tremendous losses to people's lives and property. The ecological environment of the ravaged areas was seriously damaged, and the subsequent spread of plague, oncomelania, and schistosomiasis caused numerous deaths and devastating emotional trauma. In the floods that occurred in 1931 and 1935, for instance, a death toll of over 100,000 people was reported each time. The 1954 flood, which was combated with great success under the party's and the government's leadership, still caused a death toll of over 30,000. When severe flooding occurred in the lower Changjiang River in 1983, the floodwater was diverted to Dengjia Lake. Because of the widespread occurrence of black rats (*heixianjishu*), which caused the dispersion of haemorrhagic fever, in Maliang town alone, a district of merely 40,000 people, 3,600 contracted haemorrhagic fever, more than 210 died, and 21,500 suffered from dysentery, which also caused a significant number of deaths. The flood also led to widespread occurrence of oncomelania. After the 1983 flood, the area affected by oncomelania in Jiangtan, Nanjing Municipality, jumped from 2,000 *mu* to 30,000 *mu*, over ten times more than before the flood. Also, a large number of people were infected by exposure to floodwater during flood prevention and diversion, causing the incidence of acute schistosomiasis to reach 1,000 annually in recent years. In 1954 when the floodwater was diverted to the Jingjiang flood diversion region, the number of people transferred to unflooded areas increased so dramatically that the resettlement areas had to accommodate 9,000 persons per km^2. Because of poor living conditions and widespread diseases, the death rate reached as high as 15 per 1,000. After the flood, people returned to their homes and suffered again from deteriorated living conditions and tremendous emotional trauma. A significant number died for this reason. It is evident that flood damage has a very serious impact on the ecological environment.

In the recent period of over three decades, embankments in the middle and lower reaches of the Changjiang River were heightened and reinforced. A total of 3 billion m^3 of earthwork and 65 million m^3 of stonework has been completed. Moreover, both flood diversion projects in Jingjiang and Dujiatai and large-scale reservoirs in Danjiangkou, Taxi, and Talin were constructed. However, the danger of floods in the middle and lower reaches of the Changjiang River remains serious. The embankments in various places can withstand floods that occur once every five to twenty years. Should a flood exceeding the above severity occur, diversion of floodwaters would be needed, but it would cause serious losses. The situation is most serious in the lowlands north and south of the Jingjiang section, which has an area of 40,000 km^2. During flood periods, therefore, the floodwater will often rise to a level more than 10 m higher than the surrounding level of the north bank. If an especially severe flood similar to the 1870 flood occurs, there will be no safe means of diverting the floodwater. The unstable parts of the embankments could breach, and it will cause a catastrophic number of casualties and huge economic losses. As well, navigation in the middle Changjiang River might be cut off, disrupting the economic development of the area. The recurrence of a flood like the one in 1954, even assuming that the floodwater is diverted as planned, will cause the inundation of 14 million mu of farmland, emergency displacement of 7 million people, the loss of 18 billion yuan for floodwater diversion and storage, and the loss of 3 billion yuan for flood prevention and industrial and transportation losses in cities. The total direct loss incurred is 21 billion yuan. This amount, however, does not include the negative impact on various sectors of the national economy and on the ecological environment, some of which cannot be evaluated in monetary terms. Once flooded, the lowlands in the middle and lower reaches will be inundated for a long duration. Often the damage inflicted by a single flood on the human ecosystem and the natural environment will take more than a few years to repair itself.

According to the TGP scheme, with a normal water storage level of 175 m, the flood prevention reservoir will have a capacity of 22.15 billion m^3. The project will ensure that the Jingjiang embankments will be protected from the hundred-year flood, and thus a massive loss of life can be prevented. For floods of a higher frequency, the project will reduce the risk of the Jingjiang and Chenglingji regions from being inundated, will maintain the water level in Shashi at 44.50 m, and will not require the diversion of floodwater into the Jingjiang flood diversion region. Meanwhile, the project will also create favorable conditions for constructing regulatory gates at three inflow sites of Lake Dongting, which will help develop the Dongting region. The TGP, in coordination with the construction of embankments as well as flood storage and diversion projects, will mitigate the threat of severe flooding in the middle and lower reaches and reduce the risks associated with breaching of the Jingjiang embankments during floods. This will also reduce the enormous labor and financial burden for embankment construction and flood prevention and the frequency of diverting floodwater into the flood diversion regions. The social and environmental problems created by flood prevention and diversion will thus be greatly alleviated. Moreover, the project will help control the widespread occurrence of schistosomiasis, reducing the loss of lives and property and the psychological threat created by flooding in the lower Changjiang valley. Hence, it will improve the region's production and living conditions as well as impart a sense of security to the people. Besides, siltation in Lake Dongting will be reduced, the lake's life span lengthened, the regulatory and storage capacities expanded, and the lacustrine ecology improved. It is evident that flood prevention can bring about enormous social and environmental benefits.

4.2. Large supplies of clean power as opposed to thermal power to avoid environmental pollution

Hydropower is a renewable and clean energy resource. The construction of dams for power generation is an important measure of utilizing water energy. The Changjiang River has an enormous flow volume and potential hydropower resource. Geographically well-situated, the proposed Three Gorges reservoir is located at the heartland of the river basin in Central China. According to the 175-m scheme, the total generator capacity of the Three Gorges power station will be 17.68 million kW, and the annual average power output will be 84 billion kWh. Undoubtedly, the project, which will transmit a huge amount of hydropower to energy deficient yet needy regions such as Central and East China, will significantly contribute to the development of the national economy and improvement in the environmental quality of the regions. The use of hydropower will alleviate the environmental and social problems that might otherwise be created by the use of other forms of energy. A thermal power station with an equivalent power output will require the burning of 42 million tonnes of raw coal a year, discharging after combustion 2 million tonnes of carbon dioxide, 10,000 tonnes of carbon monoxide, 370,000 tonnes of nitrogen oxides, and over 12 million tonnes of ash. In addition to the environmental problems caused and capital investment required by the exploitation of coal mines, the construction of a transport system, and shipping arrangements, large amounts of water resources are consumed. The discharge of the water used for cooling also causes thermal pollution in part of the water body. All these poisonous, harmful substances, when discharged without treatment, will have severe adverse effects on the ecological environment. In contrast, the same amount of power can be generated at the Three Gorges without causing the pollution problems mentioned above because it is a source of clean energy.

4.3. Effective improvements to navigation in the reservoir area and downstream of the dam site

Navigation in the Changjiang River plays a vital role in China's inland navigation. Though Changjiang is the third largest river in the world, the development of its transport capacity has not lived up to the status it deserves. Its annual transport capacity is far smaller than comparable rivers in other parts of the world. This is because, except for the Gezhouba area, the river channels have remained unimproved, especially the Chuanjiang and Jingjiang sections, where navigability is extremely poor.

The section between Yichang and Chongqing is referred to as "Chuanjiang" and is 660 km in length. It passes through high mountains, hills, and deep gorges. The current is swift, with reefs and shoals appearing one after another. There are a total of 139 shoals and 46 stretches of one-way controlled navigation. Tourists, local and foreign alike, are carried away by the enchanting scenery of the Three Gorges on the Chuanjiang section. However, the reefs and shoals, roaring waves, and rapid currents have also made the Three Gorges a perilous course for shipping. Here, many shipwrecks occurred and people were drowned in recent history. The situation is especially serious in the upper reaches, where the transportation capacity is small, leading to long passage time and high costs. The transportation efficiency per unit power is merely one-tenth that of the middle and lower reaches, while the transportation cost is more than double. The Jingjiang section downstream of the dam site is a meandering channel with frequent formation and erosion of bars. Channel changes are highly unpredictable. During the dry season, exposure of bar deposits is frequent and unobstructed navigation is not ensured. All these have restricted the development of navigation and the tourism industry and fail to instill a sense of security.

After construction of the Three Gorges reservoir, the backwaters during the dry season will reach as far upstream as Chongqing, submerging all the rapids and shoals on the Chuanjiang section. Navigability in the reservoir area will be greatly improved. Meanwhile, the discharge in the middle and lower reaches downstream of the dam site will increase, channels will be deepened, and the Jingjiang waterway will be improved as well. In this way, the Changjiang River, a "golden waterway" traversing from east to west in China, will be of strategic importance in the continued economic development, social stability, and prosperity of the whole river basin.

4.4. Aquatic environment for fish species and marine organisms will be expanded in favor of fisheries

After construction of the dam, the ecological environment of the river channel in the reservoir area, which is characterized by numerous shoals and rapids, will be changed. The water surface area will nearly triple. The reservoir waters will become more fertile and cleaner, because some of the organic matter and nutrients upstream will be trapped in the reservoir area. These new conditions will favor the reproduction and growth of fish species and organisms on which fish can feed. In winter, an increase in discharge downstream of the dam and in the water levels of the rivers and lakes will provide more favorable conditions for fish species to survive. However, the waters in the main river channel in the Three Gorges reservoir area will be too deep for the development of favorable fishing grounds. So emphasis should be given to the study and development of fishing grounds in sheltered bay areas as well as of new fish species adaptable to the new environmental conditions. In summary, the development of fisheries in the reservoir area and downstream of the dam is very promising.

4.5. A more charming environment for tourists will be created

The Three Gorges Reservoir area was the territory of the ancient Ba and Chu kingdoms for a very long time. The area abounds in scenic spots and historic relics, and the Three Gorges is especially famous worldwide for its imposing and precipitous, yet serene and charming scenery. After construction of the dam, "a peaceful lake will emerge at the deep gorges." The scenery of lakes and mountains, together with the grand reservoir project, will add luster to the wonderful scene at the Three Gorges, which will become a famous scenic spot. Also, the local people will be provided with more employment opportunities in the tertiary industry.

Moreover, the water quality and water supply in the middle and lower reaches will be improved during the dry season.

In short, the TGP will foster advantages and eliminate disadvantages. Its various functions will play an important role in promoting socioeconomic development and in improving the environment.

5. Comprehensive assessment and analysis of the environmental impact of the TGP

The TGP involves numerous factors that are interrelated in a complex way. Many of the environmental factors are difficult to quantify at present. To arrive at a comprehensive assessment, we have explored different methods and attempted to quantify and assess the impact of various complex environmental factors. The reliability of the results, however, depends in large part on the experience and perception of the participants in the assessment. The results, therefore, can hardly reflect reality in an objective and accurate manner.

In conducting the systematic project assessment, we consulted the method proposed by the International Committee on Dam Construction and Environment but with reference to the actual conditions of the TGP. In conducting the comprehensive analysis, we have followed several principles and methods: (1) A socioeconomic system was included, which took into account the environmental impact of important project functions such as flood prevention and power generation. (2) The whole basin was treated as a megasystem, and, where appropriate, quantitative and qualitative assessment at both the macro and micro scales were integrated. (3) For negative impacts, scenarios both with and without remedial measures were considered.

The following are the results of the comprehensive analysis of the ecological and environmental impact of the TGP:

1. The TGP has both positive and negative impacts on the environment. The overall impact is that the positive effects will outweigh the negative ones. The project will bring not only enormous economic benefits because of flood prevention and power generation, but also social and environmental benefits.

2. The TGP will affect in varying degrees the ecological environment of the reservoir area, the middle and lower Changjiang River, and the estuary. However, the impact will be most felt in the reservoir area, and among the many impacts, environmental problems created by flooding and resettlement will be the most

serious. By proper implementation of the tasks of comprehensive planning for the reservoir area, resource-exploitative migration, effective measures and policies, active development of productive forces, and establishment of a new human ecosystem, we believe, the population carrying capacity of the reservoir area will be acceptable.

The TGP will reduce flood disasters in the middle and lower reaches of the Changjiang River, save people's lives and property, and protect the ecological environment. Such positive impacts are highly significant.

The project has both positive and negative impacts on the ecological environment of the river estuary. The adverse effect occurs in October, when the reservoir is impounded, and the decrease in the regulated flow downstream will cause environmental problems. However, this will not lead to serious consequences.

3. In the reservoir area, the adverse effect of inundation caused by impoundment on the biophysical and the social environments will be the most serious. However, except for a few irrevocable and unrecoverable environmental components (for example, land, cultural relics, and landscape resources), most of the adverse effects can be avoided or alleviated through remedial measures. Some of them are as follows:

The life of the reservoir can be substantially lengthened if the siltation problem is controlled by following the principle of "retaining clear water and releasing turbid water" and by other auxiliary measures.

Though problems of environmental geology such as earthquakes, rockfalls, and landslides induced by the construction will not pose a threat to the safety of the dam, monitoring and forecasting work should be strengthened.

Cultural and historic relics should be saved by measures such as relocation, preservation, and reproduction.

Precious organic resources and rare animals and plants should be saved from extinction through transplanting, artificial breeding, and conservation.

Based on the principles of suiting measures to local conditions, utilizing favorable factors and averting harmful factors, and overcoming drawbacks and developing strong points, the pattern of land utilization should be adjusted, the potential of the mountain area and macro agriculture developed, and the agroecology improved.

Soil erosion should be prevented through reforestation in the reservoir area, and soil and water conservation.

In building new cities and towns and industrial and mining enterprises, and in exploring resources, pollution control should be incorporated into the planning, construction, and operation process of the projects.

The cleaning up of the reservoir, hygiene, and epidemic prevention should be strengthened to forestall infectious and epidemic diseases.

Many of the remedial measures are important components of the development plan in the reservoir area and are closely related to the resettlement project. If we work according to the natural and economic laws, and if we can coordinate the construction and development of the reservoir area with the protection of the ecological environment, we should be able to establish a new ecological balance that will ensure the maintenance of a benign cycle for the ecological environment of the reservoir area.

4

The Three Gorges Project Should Not Go Ahead in the Short Term

Economic Construction Group of the Chinese People's Political Consultative Committee

DURING the third session of the Sixth National Political Consultative Committee, many committee members expressed diverging opinions about the Three Gorges Project (TGP) and suggested that a field investigation of the project be conducted. In accordance with the opinions and demands of the members, the Economic Construction Group of the committee organized an investigative group led by its director, Sun Yueqi. Deputy leaders of this group were Lin Hua, deputy director of the Economic Construction Group and former vice-minister of the State Planning Commission; Xu Chi, adviser of the State Economic Commission; and Wang Xingrang, former vice-minister of commerce. Other members of the group were drawn from commission members and experts from other quarters. A thirty-eight-day field investigation was conducted from May 30 to July 6 in localities including Chengdu, Chongqing, Wanxian, Zigui, Yichang, Shashi, and Wuhan. Following the principle of "observe, listen,

---

Economic Construction Group of the Chinese People's Political Consultative Committee, "Sanxia gongcheng jinqi bunengshang." *Diqu fazhan zhanlue yanjiu* (Research on regional development strategies) 9, 3 (1987): 89–92. Translated by Ngan-chu Liu.

question, and record," the investigative group convened more than forty forums as well as on-site investigations. A total of over four hundred people attended the forums. They included high-ranking officials from ministries and bureaus of economic construction, transportation and communication, water resources and electric power, geology, seismology, meteorology, and hydrology; experts and scholars specializing in related fields; and members from the national and local political consultative committees. The investigative group consulted opinions from all quarters on the question of the TGP.

The main objective of the investigation was to explore the desirability, feasibility, and timing of the TGP. Based on our investigations, our basic conclusion is that the TGP should not go ahead in the short term, or at least not during the Seventh Five-Year Plan period. The reasons are explained below.

1. Investment—not 20 billion but 60 billion yuan

According to the Planning Office of the Changjiang River Basin (POCRB), the 150-m scheme [referring to a normal pool level of 150 m—ed.] of the TGP will involve a total investment of 20 billion yuan. However, our investigation shows that the project will in fact cost over 60 billion yuan. Moreover, the construction of the TGP would lead to a series of induced investments. For instance, the installed generator capacity of the project will be 13 million kW, but the guaranteed power output is only 3 million kW. Thus, when demand exceeds the guaranteed level, linkage of the project to the national power grid alone is insufficient to ensure a reliable power supply to the industrial and agricultural sectors. Thermal power plants of a significant size will have to be built. In addition, according to the 150-m scheme, 300,000 residents will have to be temporarily resettled and a large number of factories and houses inundated if floodwaters were stored between the flood prevention levels of 150 m and 170 m. Also, the requirement for channel maintenance as a result of siltation is another problem that needs to be resolved through state investment. Thus, this entire series of induced investments will affect the overall performance of the program of four modernizations and should not be neglected.

2. Flood prevention—the project would not solve the problem of flooding in the middle and lower reaches of the Changjiang River; it would increase the severity of flood damage in the upper reaches

According to recent surveys, the following are the major sources of floodwaters ravaging the Changjiang River: (1) The Chuanjiang section [between Yibin and Yichang]; (2) the Hanjiang River; and (3) the Xiangjiang, Zishui, Yuanshui, and Lishui rivers. The Chuanjiang floodwaters alone are insufficient to cause disasters in the middle and lower reaches. In 1984, for instance, especially severe flooding occurred in Sichuan and caused heavy losses there, but the middle and lower reaches of the Changjiang River remained intact. The 1954 flood in Sichuan, on the other hand, was not large, but because severe rainstorms occurred in the river basins of the Hanjiang, Xiangjiang, Zishui, Yuanshui, and Lishui rivers, the middle and lower reaches suffered from disastrous flooding. As the TGP dam will be located upstream from the confluences of the Changjiang River and the Hanjiang, Xiangjiang, Zishui, Yuanshui, and Lishui rivers, it will only impound floodwaters from the Chuanjiang section but not from the other two major sources. Moreover, the storage capacity of the reservoir for flood prevention is limited. The POCRB estimates that if a severe flood like the one in 1954 should recur, the Three Gorges reservoir could only detain 9.3 billion m^3 out of a total 70 billion m^3 of floodwater. The remaining amount will not be detainable. Disastrous flooding would also occur if the floodwaters from the Changjiang and Hanjiang rivers or from the Xiangjiang, Zishui, Yuanshui, and Lishui rivers converge. In fact, following construction of the TGP, not only must the Jingjiang embankments continue to play an important role, the flood diversion region in the vicinity will still have to be used as well.

The TGP not only would play a minor role in preventing floods in the middle and lower reaches of the Changjiang River, it would also increase the severity of flood damage in the upper reaches. Flood drainage is already a serious problem in the Sichuan Basin, where the Changjiang River is the only channel for surface drainage. After construction, the severity of flood damage in the upper reaches will be exacerbated whether the 150-m or the 180-m scheme is adopted. Chongqing Municipality has expressed a preference for the 180-m scheme. For this scheme the Sichuan Academy of Sciences has suggested that a hundred-year flood will raise the water level in the middle of the reservoir to 190 m. According to the laws of fluid mechanics, the water level at Chongqing will exceed 210 m, while those in localities such as Jiangjin and Hechuan will exceed 220 m, and backwaters will reach the Sichuan Basin. In the Jialing River basin and the relatively open area between Hechuan and Tongliang, not only will the water levels rise, but the duration of flooding will also increase. In the Chongqing-Jiangjin-Hechuan area, which has a dense population and a heavy industrial concentration, the losses due to inundation should be investigated. If a major dam requires tens of billions of yuan to construct, but it is only capable of transferring flood damages from the lower to the upper reaches—or in other words, if flood damage in the middle and lower reaches cannot be prevented while flood damage in the upper reaches is exacerbated—then the cost-effectiveness of such a project should definitely be investigated.

3. Siltation—an unresolved problem

The POCRB has analyzed the siltation problem in the Changjiang River on the basis of its mean annual sediment load. However, because of the destruction of vegetation in the upper reaches, serious soil erosion has occurred, and sediment load in the river has markedly increased in recent years. For instance, the mean annual sediment load has increased from 520 million tonnes in 1981 to 680 million tonnes in 1984. Such a large amount of sediment, though less than that in the Huanghe River, is still a serious problem given that navigation is an important component of the TGP.

According to a brief prepared by the POCRB, methods such as "releasing turbid water and storing clear water" and the use of vents for discharging sediment would solve the problem of siltation. This view is hardly convincing. This is because the TGP is a multi-purpose project designed for flood prevention, power generation, and navigation. These objectives will pose constraints on the operation of releasing turbid water and storing clear water. During the dry season, for instance, when the water is the clearest, power generation and navigation in the middle and lower reaches would require more water to be released. However, during the rainy season, when the water is most turbid, flood prevention policies would require the water to be stored. The problem would be even greater during periods of flooding, when the sediment load is the highest. Flood peaks tend to occur in succession, and each time floodwater is stored, siltation would occur. In time, a large amount of sediment will accumulate. It is certainly possible to vent sediment in the vicinity of the big dam, but it is unlikely from as far as Fuling and Changshou, which are located 400 to 600 km upstream.

According to our survey of the results of experimentation and the present field situation, there is no doubt that navigation can be impeded by siltation. We do not believe that the problem of siltation in the Three Gorges Reservoir has been solved. The view of the POCRB is that the problem has been resolved. This is based on the experience gained at the Gezhouba (Gezhou Dam). However, according to those involved in the on-site survey of the waterway, the siltation rate at the Gezhouba is so high that soon pebbles may get into the sluices. Clearly, it is uncertain whether the problem of siltation at the Gezhouba has been successfully resolved, and it is premature, therefore, to conclude that the problem of siltation in the TGP has been resolved through the experience at the Gezhouba. Whether this experience is applicable to the massive TGP is yet another problem deserving further investigation.

4. Navigation—more harm than good

Basically, the impact of the TGP on navigation will be to limit the carrying capacity of the Changjiang River. The design vessel capacity of the dam is 50 million tonnes per year. This is only a theoretical figure based on the assumption of standardized vessel size (3,000 tonnes each) and automated regulation (that vessels will queue up in an optimal fashion and that they will pass the dam without any delays at all). In practice, such a figure cannot be achieved. Moreover, as the dam is designed to store water for power generation, to release turbid water and store clear water, and to divert water to transport sediment, navigation would be adversely affected. Even assuming that the passage capacity of 50 million tonnes is achievable, such a permanent structure may not be sufficient to meet the needs of China's future economic development. Once completed, therefore, the big dam would become a hindrance to the economic development of southwestern China. Furthermore, in the 150-m scheme, 10,000-t vessels could not reach Chongqing, and the port of Chongqing will not be functional because of siltation in the backwater area of the reservoir. Once this happens, not only will shipping on the Changjiang River but also the economic development of the Southwest and even of the whole country be adversely affected. Therefore, the losses in navigation caused by the TGP could by no means be compensated for by benefits brought about by power generation.

The only benefit brought by the TGP to navigation in the Changjiang River would be improvements to the 300-km long waterway upstream of the big dam. The total length of the waterway from the dam site to Chongqing is 600 km, and the effects and benefits brought about by transportation improvements to only a section of the waterway are limited. On the whole, the benefits are outweighed by the adverse effects of the TGP on navigation.

Moreover, there are a number of problems of navigation associated with the construction of the TGP which are still unresolved. According to a brief prepared by the POCRB, the boat-lifting locks at the main dam will be modeled after the 300-t system at Danjiangkou. Yet, the departments concerned with navigation suggest that the Danjiangkou model is not suitable. There is, therefore, some doubt about the reliability and appropriateness of the current design of the 10,000-t boat-lifting locks at the main dam. In addition, there are many other problems which require further study. These include whether the multilevel boat locks could ensure unimpeded navigation; the effects on navigation safety when unsteady flows develop between the big dam and the Gezhouba because of the need to regulate peak flows for power generation; how serious is the problem of lowered water levels at the sluice thresholds due to downcutting of the river bed downstream from the dam; and the effects on navigation due to siltation in the channel approaching the dam upstream.

5. Power generation—high investments, long construction period, slow outputs, poor results

The 150-m scheme of the TGP has a designed total generator capacity of 13 million kW. When compared to the estimated investment of 20 billion yuan as suggested by the POCRB, the project is certainly cost effective considering the capacity of power generation alone. Our investigations show, however, that the total investment could exceed 60 billion yuan, and the unit investment will be as high as 4,600 yuan per kW. Even if interest charges are not included, the total investment will still exceed 31 billion yuan and the unit investment over 2,300 yuan per kW. When compared with an average unit investment of some 1,600 yuan per kW for several large- and medium-scale hydropower stations that are currently under construction, it is obvious that the project is not very cost-effective.

As the TGP would require a long construction period, there will be merely inputs and no outputs for more than ten years after construction begins, and the dam will not be functional until after the year 2000. Not only will it take up an enormous amount of precious construction funds but it will also, as a result, cause the abandonment of construction projects that are urgently needed for quadrupling [the gross value of industrial and agricultural output by the year 2000 to achieve the national overall objective of the four modernizations].

At present, quite a number of work units and individuals have proposed some alternative schemes. They all suggest building some large- and medium-scale hydropower stations on the tributaries in the upper-middle reaches before building the TGP. These stations generally would share the characteristics of short construction period, small investment, quick outputs, and early results. Since preliminary design already exist for some of them, construction can begin as planning for each is completed. This is perfectly in line with the principle of working according to the resources available. Objective assessment should be conducted for all the suggested alternatives.

6. Resettlement—more than ten cities to be reconstructed

At present, large variations exist in the estimated resettlement population and the costs incurred among the many parties concerned. The POCRB estimates that for the 150-m scheme, 330,000 people in the inundated zone will have to be resettled, and the cost incurred will be 3.5 billion yuan. However, the Chinese Urban Planning and Design Institute and eleven of the fourteen counties in the proposed inundated area estimate the number of residents as 860,000 (not including residents from three counties affected by the proposed inundation) and the cost at 11 billion yuan. These two sets of estimates differ widely, and according to our investigation, the latter are more realistic. Even if the estimates of the POCRB were adopted, it may be difficult to implement resettlement plans, and this may lead to serious consequences.

The Changjiang River Basin is the prime region of industrial and agricultural production in China. The cities on the banks of the Three Gorges have evolved over a long period. Some of them have a history of several hundred or even over a thousand years. The 150-m scheme of the TGP would inundate more than ten cities along the river as well as the foundations and production facilities that some one million people have been relying upon for their livelihood. Whether the construction of such a project is warranted under the present economic conditions in China requires careful consideration.

7. Safety—taking a disastrous risk

7.1. Rockfalls and landslides

The banks of the Three Gorges reservoir area are not the intact, solid rock walls they are popularly believed to be. Rather, they have a very complex and fragile geological structure. According to the on-site surveys made by Sichuan's Bureau of Geology over the last couple of years, rockfalls or landslides occurred at as many as 203 localities. In over 30 of these localities, the slides involved more than 10 million m^3 of material each. Most disturbing is the fact that these localities include Lianziyan and Xintan, both only 27 km away from the proposed Three Gorges Dam.

Lianziyan is located on the south bank of the Changjiang River. There are twelve cracks in the rocks at higher elevations. At 310 m above the river is one of the largest bodies of rock, which measures 2.5–3 million m^3 in volume and is estimated to weigh more than 7 million tonnes. According to the local old-timers, the cracks in the rocks have widened considerably in the last several decades. If these rocks should break loose and fall downslope, the largest piece alone would release over 2 billion meter-tonnes of energy, which is the level of energy released in an earthquake. It is hard to predict the consequences and the destructiveness to the project should such failures occur. What is more, rockfalls trigger landslides, which in turn can aggravate the danger of further rockfalls, and rockfalls at Lianziyan could trigger a chain reaction of landslides at Xintan.

The landslide body at Xintan, which is opposite Lianziyan

across the river, is a mass of rock debris resting on the bedrock and measures some 800 m long, 900 m high, and 70 m thick. If the mass of rock debris all crash into the Changjiang River, which is less than 300 m in breadth at this point, the river would be blocked, forming a "dam" behind the big dam and rendering the latter ineffectual.

According to geologists, the occurrence of these landslides is related to the interbedded formation of shales and limestones in the Three Gorges area. On the surface of the shales is a weathered layer which, when saturated with water, will become slippery and leads to landslipping. As historical records show, landslides usually occur after heavy rainfall. Before the recent landslide at Xintan, however, there had not been any rainstorms, but near the top of the falling rock debris, a jet of water was observed gushing several meters into the air. It was inferred that the raising of the water level due to the completion of Gezhouba had caused the cracks to be saturated with water. After the construction of the big dam at the Three Gorges, the water level will be raised by more than 100 m. Whether this will aggravate the danger of landslides in this area and what the consequences of such changes will be are questions that require further investigation.

7.2. High probability of triggering earthquakes

According to the seismological departments in Wuhan, there are now over eighty large and medium-size dams in the world that have triggered earthquakes. In some of these localities, there were small earthquakes before the construction of the dam. In some cases, earthquakes have never been recorded before. Damage caused by reservoir-induced earthquakes is severe. Therefore, the question of the TGP-triggered earthquakes should not be neglected. The consequences of such neglect could be catastrophic.

In the vicinity of the Three Gorges Dam, there are three seismic fracture zones: Xiannu Mountain, Jiuwanxi, and Tianyangping. The three zones converge right at Xintan and Lianziyan, where an earthquake measuring 5.1 on the Richter scale occurred recently. It is evident that this area is prone to induced earthquakes. As stresses in the rocks would change after construction of the reservoir, once earthquakes are triggered, not only would the dam be in danger, but landslides and rockfalls would also be induced. Therefore, to avert catastrophic damages, the question of triggered earthquakes after the reservoir is impounded and their consequences should be thoroughly scrutinized before any final decision is made. Yet, an examination of the information provided by the departments concerned shows that surveys on the geological structure at great depths either at the dam site or in the reservoir area have not been conducted.

Moreover, the big dam at the Three Gorges would become a key target of attack in times of war. Once the big dam is hit, the lives and property of the people in Wuhan and in the middle and lower reaches as well as the whole war situation would be gravely affected. This problem cannot be ignored.

In light of the above circumstances, the survey group believes that it is inappropriate to go ahead with the TGP in the short term. As this is a huge project that will attract worldwide attention, standard procedures for capital construction should be followed even more strictly. Construction work for the initial stage should be halted to avoid waste, while scientific research at this stage may continue.

Thorough assessment and comparative studies of alternative schemes should be carried out to determine the feasibility of the TGP. In the course of assessment, we should exercise democracy and listen to wide-ranging opinions from all quarters. In handling such a massive project, we should take a particularly serious attitude in consulting different views to arrive at a correct decision.

5

Some Aspects of Sedimentation at the Three Gorges Project

Qian Ning, Zhang Ren, and Chen Zhicong

SEDIMENTATION at the Three Gorges Project (TGP) has long been considered an important problem by members of the Central Committee of the Communist Party. It has also been the concern of people from all walks of life at home and abroad. Although many articles dealing with the various impacts of the TGP have been published, the discussion has not been sufficiently comprehensive and detailed. In this paper, our objective is to discuss several problems concerning sedimentation at the TGP based on the results of research conducted in the last few years. It is our hope that this discussion will contribute to the development of more detailed research on sedimentation at the TGP.

1. Will the TGP be rendered useless by sedimentation?

1.1. Adopting a rational mode of operation is the key to maintaining the reservoir's effective storage capacity

To maximize benefits brought about by the comprehensive utili-

Qian Ning, Zhang Ren, and Chen Zhicong, "Changjiang sanxia shuniu gongcheng de jige nisha wenti." *Changjiang sanxia gongcheng shengtai yu huanjing yingxiang wenji* (Symposium on the ecological and environmental impact of the Three Gorges Project), 1988, pp. 105–20. Translated by Weixue Cheng.

zation of the TGP, an appropriate storage capacity should be maintained for the reservoir for an extended period of time. Such storage capacity should not only be unaffected by sedimentation during the normal operation of the reservoir, but, when high siltation rates occur, it should also be possible to restore the appropriate storage capacity through retrogressive scouring of the sediment under low flow conditions.

The key to maintaining the reservoir's effective storage capacity is to adopt a rational mode of operation. The operating mode of "impounding clear water and discharging turbid water" adopted for the design of the Three Gorges Reservoir is the correct method which will maintain the effective storage capacity of the reservoir for an extended period. This means that the reservoir is kept at a low water level during the flood season (June–September) so that the flood water and sediment are readily discharged. Toward the end of the flood season (October), impoundment begins, and by the end of November the normal storage level is attained. This will ensure the needed water supply in the dry season. During the dry season (January–April), the storage level is gradually lowered to the minimum operating level. This is to ensure adequate power output and sufficient flow rate downstream to facilitate navigation. In the process of determining the minimum operating level in the dry season, the requirement of both power generation and navigation upstream and downstream have to be considered. In this way, we believe that the effective storage capacity of the Three Gorges Reservoir can be maintained for an extended period without the danger of it being rendered useless by siltation.

The operation mode of "impounding clear water and discharging turbid water" was proposed on the basis of experience working in reservoirs associated with rivers that have a high sediment load. Sanmenxia Reservoir is a case in point. It was completed in September 1960. During the early period, water storage was emphasized in its operation, and it was often kept at a high storage level. This has led to serious sedimentation problems. The extension of deposition upstream has adversely affected agricultural production and people's livelihood in the Guanzhong Plain area of Shanxi province. In March 1962, operation was modified by emphasizing flood detention. Because the outflow rate is limited, however, water level in the flood season could not be lowered, and sedimentation continued. By October 1964, the storage capacity (below an elevation of 330 m) was only 2.21 billion m^3, which was only 38 percent of the original 5.89 billion m^3, and this trend might continue. In view of this situation, it was decided in 1964 that the Sanmenxia Reservoir should be modified to increase its flood-discharging capacity. After the completion of two phases of this reconstruction project from 1966 to 1972, discharge capacity at a storage level of 315 m was increased from 4,000 m^3/s to 10,000 m^3/s, and the average water level in the flood season was decreased from 314 m to 298 m. As a result, not only was all the incoming sediment completely discharged, but previously accumulated sediment was also scoured, partially restoring the storage capacity of the reservoir. By August 1972, the storage capacity (below 330 m) was restored to 3.1 billion m^3, and it has been maintained since then [1]. The temporal variation of the storage capacity (below 330 m) at Sanmenxia Reservoir is shown in figure 1.

It can be concluded from the above that the storage capacity of a reservoir can be maintained provided that the reservoir is operated at a reasonably low water level during the flood season, when sediment discharge is at its highest. Even if the effective storage capacity is reduced by sedimentation, it can be restored by retrogressive scouring of the accumulated sediment provided that the low-storage-level operating mode is practiced during the flood season.

To operate at a low water level during the flood season, the reservoir must have a sufficiently large flood-discharging capacity. In this respect, the Three Gorges Reservoir (TGR) will meet the requirement. Table 1 shows the flood-discharging capacity of the TGR at different storage levels [2]. It can be seen that at a normal pool level of 145 m, the discharge capacity is 64,210 m^3/s, which is equivalent to the peak discharge of a hundred-year flood

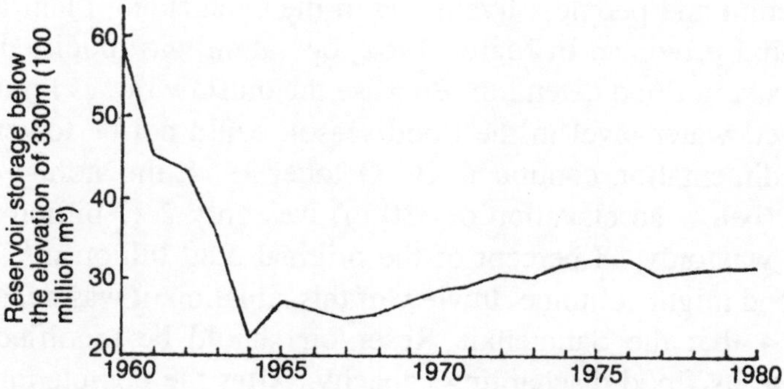

Figure 1. Changes in storage capacity of the Samenxia Reservoir below the elevation of 330 m

[3] (table 2), or 1.24 times the long-term average peak discharge measured at the Yichang hydrological station.

1.2. The TGR is a canyon-type reservoir favorable for the maintenance of effective storage capacity

It can be seen from figure 1 that the storage capacity of the Sanmenxia Reservoir cannot be completely restored by changing its mode of operation. This is because when the storage level is lowered for scouring sediment, usually only a channel is carved in the accumulated sediment. The width of this channel, known as the equilibrium channel width, depends on the inflow rate and the magnitude of flood peaks. If the river valley of the reservoir area is wider than the equilibrium channel width, then only a limited volume of storage capacity can be restored by lowering the water storage level and retrogressive scouring. On the contrary, if the valley of the reservoir area is quite narrow, or has a width similar to that of the equilibrium channel, then most of the lost storage capacity due to sedimentation can be restored. The storage capacity restored at the Sanmenxia Reservoir (SMXR) area is limited because the valley is quite wide, ranging from 400 m to 5,800 m. Thus, after more than two decades of operation, the equilibrium channel formed by retrogressive scouring was

Table 1

Flood Discharge Capacity of the Three Gorges Reservoir for the 170 m Dam Scheme

Water level (m)	Spillway (m^3/s)	Bottom sluice (m^3/s)	Power plant (m^3/s)	Sediment sluice (m^3/s)	Total (m^3/s)
135	—	39,310	17,000	2,280	58,590
140	—	41,620	17,360	2,370	61,350
145	—	43,810	17,940	2,460	64,210
150	990	45,900	18,500	2,550	67,940
155	6,500	47,890	19,050	—	73,440

only 400–600 m wide, and only the lost capacity along the channel (channel storage capacity) but not in the "flood plain" area (flood plain storage) was restored. At present, 82 percent of the SMXR's channel storage capacity below the elevation of 330 m but virtually none of the flood plain storage has been restored.

In this respect, the TGR is far superior to the SMXR. The former is a canyon-type reservoir. For most of this 600-km-long reservoir, the valley width averages less than 1,000 m. There is only a small section where the average width reaches 1,000–1,700 m after impoundment. Figure 2 shows the relationship between water stage and average river width at different sections in the TGR area, indicating that the reservoir area is quite narrow.

On the other hand, the runoff volume and the peak discharge of the Changjiang (Yangtze) River is much greater than that of the Huanghe (Yellow) River. Therefore, when the TGR reaches an equilibrium level of sedimentation, its equilibrium channel width will be greater than that of the SMXR. Data collected in the upper Jingjiang River, which is located in the middle reaches of the Changjiang River and which has similar hydrological conditions as the TGR, can be used to forecast the equilibrium channel width of the TGR area. Table 3 is an attempt to compare the runoff and flood flow characteristics of the TGR area and the upper Jingjiang section. Because the data show that little difference exists in these characteristics, it can be expected that when

Table 2

Frequency of Peak Discharge at Yichang Hydrological Station

Peak discharge (m^3/s)	Return period (year)	Frequency (%)
82,000	100	1
77,500	50	2
71,400	20	5
66,200	10	10

the equilibrium level of sedimentation is reached in the TGR, the process of sediment transport will be similar to that of the upper Jingjiang, and therefore the equilibrium channel width will also be similar. According to data from the Hydrology Bureau of the Office of the Changjiang River, the width of the upper Jingjiang section at bankfull discharge is 1,320 m [3]. It is estimated that the equilibrium channel width of the TGR area will be in this range if there are no bank constrictions.

The equilibrium channel width of the TGR can be estimated in another way. In 1985 we performed experiments on a hardware model of the Lanzhuba river section in terms of the 150-m scheme of the TGR [6]. The Lanzhuba section, which is 25 km long, is located between Zhongxian and Fengdu, at the upper end of the perennial backwater zone of the proposed reservoir. The river valley width is 800–1,700 m. After construction of the TGR, the water level will rise by 8–13 m during the flood season, causing tremendous sedimentation and resculpturing of the river bed in this section. After eighty years of operation of the reservoir, sedimentation will approach an equilibrium state. Profiles of the accumulated sediment at different locations of this river section are shown in figure 3. It can be seen that when the storage level exceeds 153 m (with a corresponding discharge of about 30,000 m^3/s), no bar deposits will be found where valley width is less than 1,200 m. Where the valley width is 1,700 m and the channel width is nearly 1,300 m, a bar 400 m wide is formed on the north bank. Thus, the equilibrium channel width of the TGR will be about 1,300 m if there are no bank constrictions.

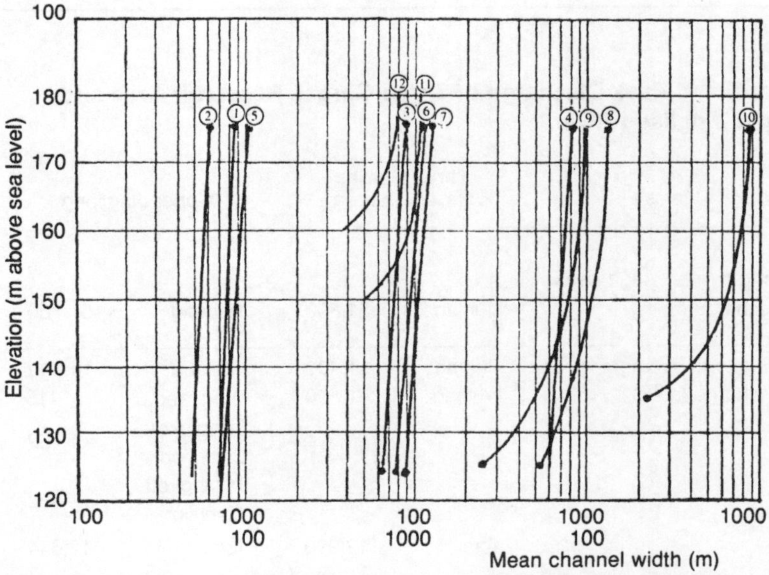

1 Dam site-Badong 2 Badong-Wushan 3 Wushan-Fengjie 4 Fengjie-Yunyang 5 Yunyang-Wanxian 6 Wanxian-Shibaozhai 7 Shibaozhai-Zhongxian 8 Zhongxian-Fengdu 9 Fengdu-Fuling 10 Fuling-Changshou 11 Changshou-Chongqing Customs 12 Chongqing Customs-CS117

Figure 2. Relationship between water storage level and mean channel width at different river reaches of the Three Gorges Reservoir

To sum up, the equilibrium channel width of the TGR area will be greater than the valley width. Therefore, as long as the operating mode of "impounding clear water and discharging turbid water" is practiced, the designed storage capacity of the TGR can be maintained in most cases. Even when floods occur simultaneously in the upper, middle, and lower reaches, which will result in severe sedimentation due to a long period of storage, most of the storage capacity can still be restored provided that during the falling stage the reservoir is operated at a low water level. The loss of storage capacity due to sedimentation in floodplain areas will not be a significant problem in the TGR.

Adopting "impounding clear water and discharging turbid water" as the mode of operation to maintain the storage capacity of the canyon-type reservoir is a feasible method. It has proven to

Table 3

Runoff and Peak Discharge of Three Gorges Reservoir Area and Upper Jingjiang River

Items	Three Gorges Reservoir area		Upper Jingjiang	
Long-term average peak discharge (m^3/s)	Cuntan	51,800	Yichang	51,700
Long-term average discharge (m^3/s)	Cuntan	11,400	Yichang	14,300
	Wulong	1,590	Qingjiang	+18
			Songzikou Diversion	−1,452
			Taipingkou Diversion	−622
	Total	12,990	Total	12,644

be effective in other rivers carrying a high sediment load [7]. For instance, the Naodehai Reservoir has been in operation for thirty-five years, and 77 percent, or 130 million m^3, of its original storage capacity (168 million m^3) is still available. Moreover, only 12 percent, or 1 million m^3, of the original storage capacity of 8.6 million m^3 at the Heisonglin Reservoir has been lost after twelve years of operation. Of course, these reservoirs are much smaller in scale than the TGR, but the principle and method designed to preserve the storage capacity of these reservoirs are similar and can be considered for application to the TGR.

2. Will the construction of the TGP render the port of Chongqing useless because of siltation?

2.1. Why is it feasible to allow the port of Chongqing to be located within the backwater zone?

In any regulating-type reservoir, there is a backwater zone where the water level is fluctuating. This is known as the fluctuating

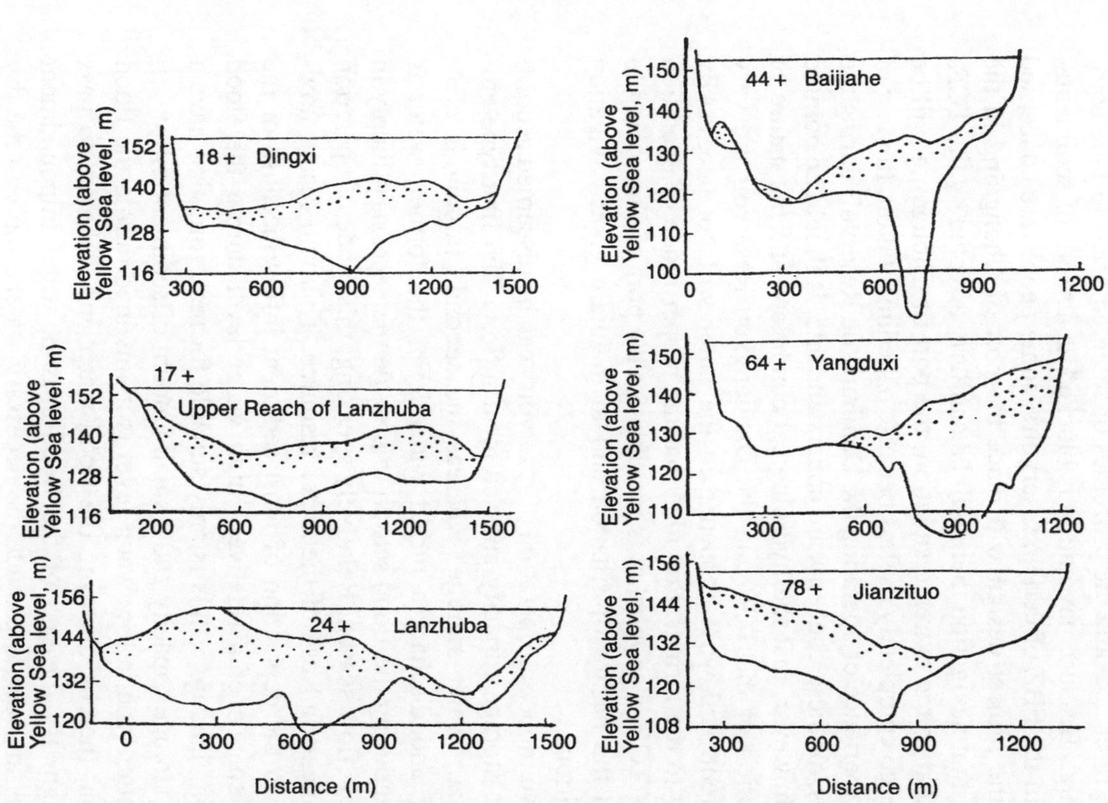

Figure 3. Predicted alluviation at different cross-sections of the Lanzhuba river reach after 80 years of operation of the reservoir
Note: Elevation at Wusong = level of the Yellow Sea + 1.8 m

backwater zone (FBZ). In this zone, water level changes frequently, alternating processes of erosion and sedimentation occur at the river bed, and the processes of sediment transport are complicated. Because navigation is blocked as a result of sedimentation in the FBZ of certain reservoirs, some people are concerned that the project design will place the port of Chongqing in the FBZ, that navigation there will be similarly affected by the TGR, and that in the extreme situation, the port of Chongqing will be rendered completely useless because of sedimentation. In view of the experience of operating the Danjiangkou Reservoir during the 1970s, all these concerns are understandable. It should be pointed out, however, that the hydrological conditions and the nature of operation of the TGR and the Danjiangkou Reservoir are very different, and what happened in the latter will not necessarily apply to the former. Provided that the proper mode of operation of the TGR is adopted, it is possible for the port of Chongqing to avoid the hazard of siltation and to maintain good navigation conditions.

Two modes of operation of reservoirs can be discerned according to storage capacity and sediment inflow. The first type emphasizes water storage, whereas the second "impounds clear water and discharges turbid water." For the former, water is stored during the flood season to provide the necessary supply in the low flow season. Relatively speaking, this mode of operation will make full use of the water resource. This will result, however, in the deposition of most sediment inflow because of the maintenance of a relatively high water level during the flood season. Therefore, this is only suitable for reservoirs that have a large storage capacity and a small amount of sediment inflow. In the latter type, a low water level is maintained during the flood season, thereby allowing the flood water and sediment to pass through. By the end of the flood season, water impoundment begins in order to provide the necessary water supply in the dry season. In this mode of operation, the water resource during the flood season is not completely utilized. However, it can reduce sedimentation and is favorable for prolonging the life of the reservoir and reaping the long-term benefits of the reservoir. Furthermore, this mode of operation can provide storage capacity for flood control. While it is true that there is always a fluctuating backwater zone in the reservoir area no matter what mode of operation is employed, the fluctuations in water level and the processes of erosion and sedimentation within the FBZ for the two cases mentioned above are entirely different.

Both the Tabora Reservoir [*sic*] in India and the Danjiangkou Reservoir in China employ the water-impounding operating mode (fig. 4). The figure shows that during the flood season, most sediment enters the reservoir when the water level is high and when the FBZ exhibits the characteristics of a reservoir. Thus a large amount of sediment is deposited in the FBZ, forming a large and contiguous body. When the low flow season begins, the water level gradually recedes, and channel sediments that are exposed are scoured and transported downstream. Because there is a large quantity of sediment to be eroded and only limited discharge is available (e.g., during the low flow season, the unit discharge of the Danjiangkou Reservoir is less than 2 m^3/s), the scouring flow readily becomes saturated with sediment, and part of the sediment is transported in the form of bed load. In straight and broad river reaches, braiding occurs, which is typically unstable, with variable discharge and migrating main channels, thus resulting in the blockage of navigation. The "iron-plate-like shoal" that occurred in the Liaowa river section of the Danjiangkou Reservoir in the 1970s was in fact deltaic deposits due in part to the accumulation of bedload material when the main channel migrates.

Under the operating mode of "impounding clear water and discharging turbid water," the characteristics of the FBZ are quite different. Because the water level in the flood season is kept low, the FBZ maintains the characteristics of a river channel, and great amounts of sediment are transported directly to the perennial backwater zone. After reaching an equilibrium in reservoir sedimentation, deltaic types of sediment will accumulate, with a thicker deposit downstream and with most of the deposition oc-

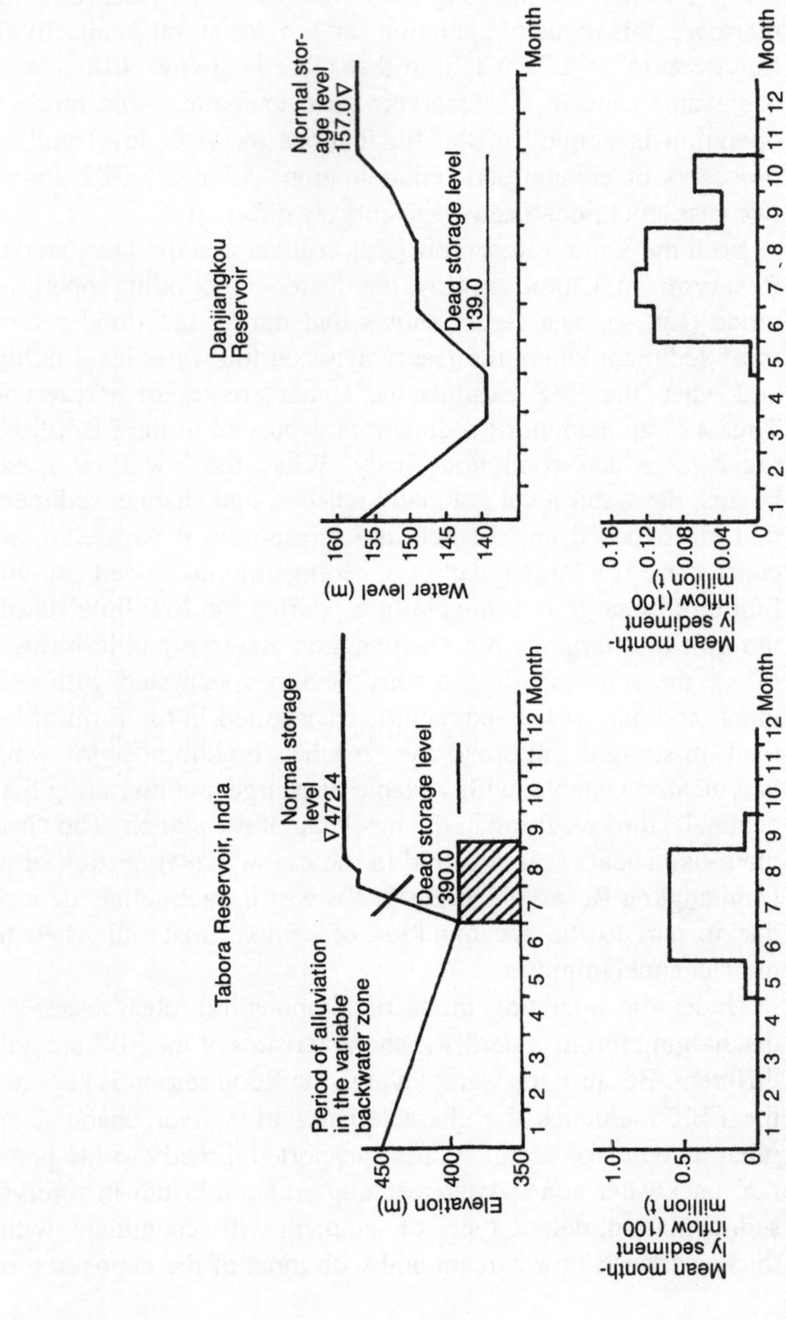

Tabora Reservoir, India
Period of alluviation in the variable backwater zone
Normal storage level
∇472.4
Dead storage level
390.0
Elevation (m)
450
400
350
1 2 3 4 5 6 7 8 9 10 11 12 Month
Mean monthly sediment inflow (100 million t)
1.0
0.5
0
1 2 3 4 5 6 7 8 9 10 11 12 Month
Danjiangkou Reservoir
Normal storage level
157.0∇
Dead storage level
139.0
Water level (m)
160
155
150
145
140
Month
Mean monthly sediment inflow (100 million t)
0.16
0.12
0.08
0.04
0
1 2 3 4 5 6 7 8 9 10 11 12 Month

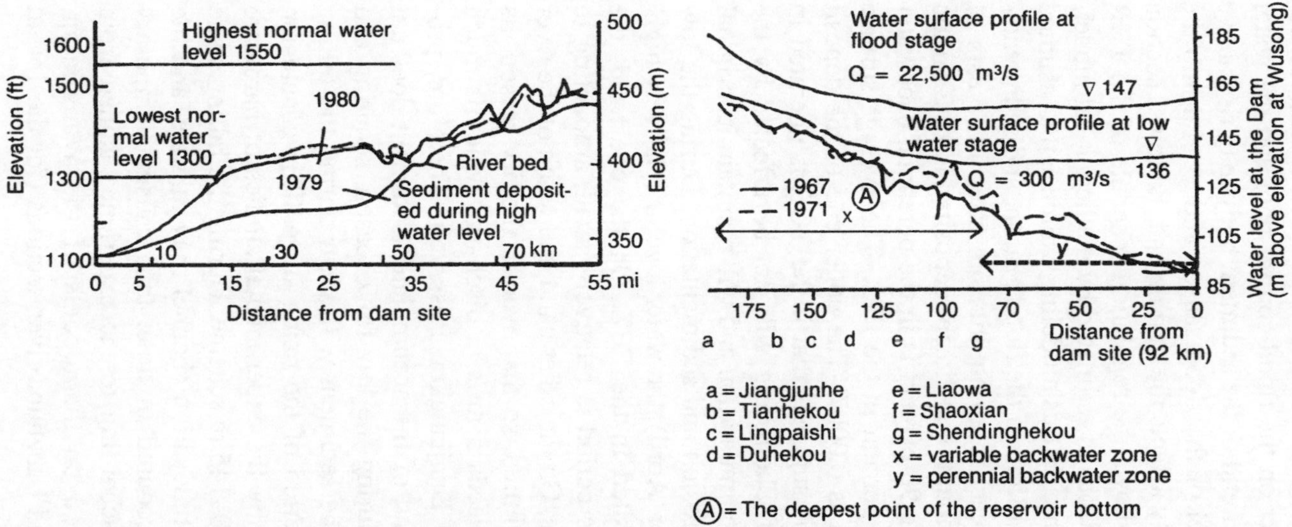

Figure 4. The regulating operation of the impounding reservoir and the profile of sediment

curring below the limiting water level of the flood season. Therefore, it has little impact on the flood control pool and the water storage pool. The upper end of the sediment body may rise above the limiting water level of the flood season and extend into the FBZ, but usually it only affects the middle and the lower section of the FBZ. Figure 5 shows the regulating schedule of the TGR and the profile of the deposits estimated by mathematical modeling for the 150-m scheme. In this modeling exercise, the assumed limiting water level is 135 m in the flood season, and sedimentation in the reservoir reaches an equilibrium after 140 years of operation. The total amount of sediment deposited is 12.2 billion m^3, of which more than 90 percent is in the perennial backwater zone and less than 10 percent in the FBZ. The average surface gradient of the deposit is 0.0007. The upper end of the deposit lies below Cuntan. This suggests that if the TGR is operated in this mode, the port of Chongqing will not be affected by the hazard of cumulative sedimentation and will maintain its original equilibrium between erosion and sedimentation. During the low flow season (November–April), the water level is relatively high, and sediment is deposited in the FBZ. On the one hand, the sediment inflow in this period is relatively small, amounting to about 10 million tonnes. On the other hand, the width of the river valley in the FBZ is limited, so that even in the dry season, its unit discharge still exceeds 5 m^3/s, a discharge which is much larger than that of the Danjiangkou Reservoir. Therefore, provided that the upper end of the accumulated sediment does not extend into the FBZ during the low flow season, scouring and complete removal of the sediment will occur during the falling stage and the initial period of the rising stage. This conclusion has also been verified by the experimental studies of hardware models for the 170–160–145 m scheme. During the falling stage in the dry season, the navigation channel may migrate, and the deposits may not be scoured in time, thus causing temporary blockages. However, according to the experience gained from experiments performed on the models of the Wangjiatan, Qingyanzi, Siguaqi, and Lanzhuba river sections (for the 150-m

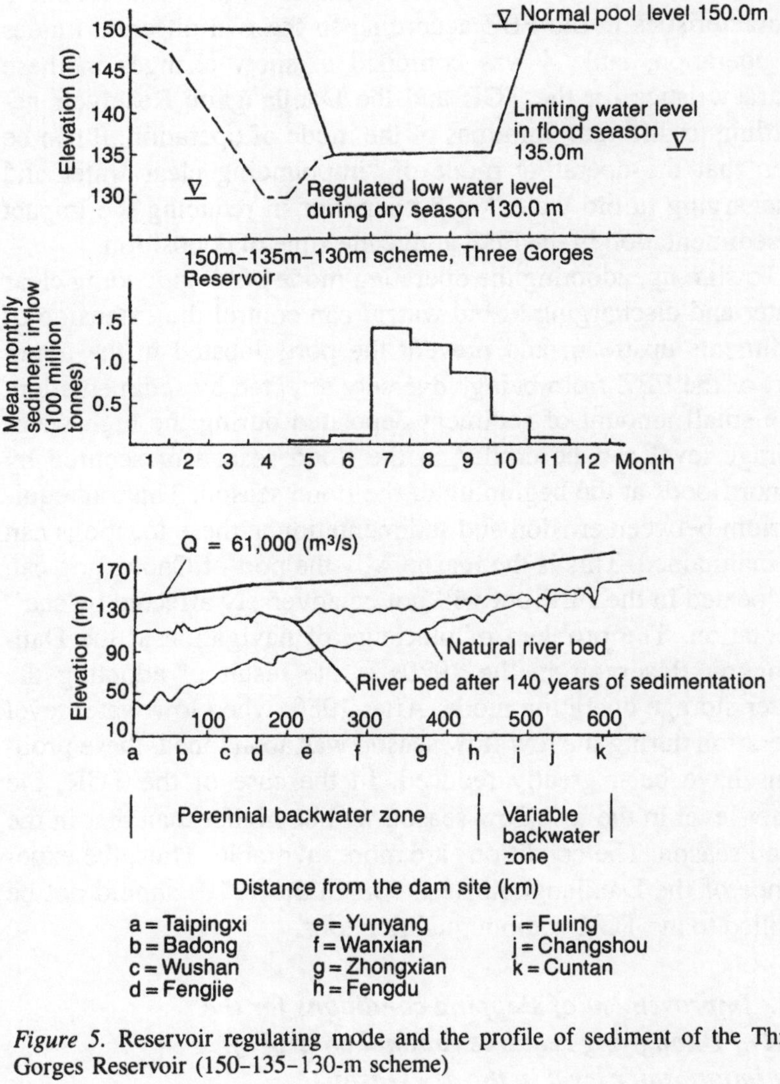

Figure 5. Reservoir regulating mode and the profile of sediment of the Three Gorges Reservoir (150–135–130-m scheme)

scheme), all these problems can be solved by dredging and other treatment.

To demonstrate the differences in hydrological and sediment characteristics in the FBZ according to the two different modes of operation, table 4 was compiled to show changes in these characteristics for the TGR and the Danjiangkou Reservoir according to different durations of the mode of operation. It can be seen that the operating mode of "impounding clear water and discharging turbid water" is far superior in reducing the impact of sedimentation in the FBZ above the zone of deposition.

To sum up, adopting the operating mode of "impounding clear water and discharging turbid water" can control the extension of sediments upstream and prevent the ports located in the upper part of the FBZ from being adversely affected by sedimentation. The small amount of sediment deposited during the high water storage level can be eroded in the flood season or scoured by minor floods at the beginning of the flood season. Thus, an equilibrium between erosion and sedimentation at these locations can be maintained. This is the reason why the port of Chongqing can be located in the FBZ but will not be adversely affected by sedimentation. The problem of blockage of navigation at the Danjiangkou Reservoir in the 1970s is the result of adopting the water storage operating mode. After 1980, when low water level operation during the low flow season was abandoned, these problems have been greatly reduced. In the case of the TGR, the water level in the low flow season will be higher than that in the flood season. The conditions are more favorable. Thus, the experience of the Danjiangkou Reservoir in the 1970s should not be applied to the TGR without qualifications.

2.2. Improvement of shipping conditions for the port of Chongqing requires rational control of minimum water level in the dry season

The low water level operation in the flood season can only protect the port basin from the adverse effects of sedimentation and

Table 4

Comparison of Hydrologic and Sediment Characteristics in the Variable Backwater Zone Between Three Gorges Reservoir and Danjiangkou Reservoir

Level of operation	Characteristics: Item	Characteristics: Unit	Three Gorges Reservoir	Danjiangkou Reservoir
High-water-level operation	Season		Low-flow season November-April	Flood season June-October Low-flow season November-February
	Average discharge	(m^3/s)	4,457	Flood season 1,167 Low-flow season 270
	Sediment inflow	(100 million tonnes)	0.11	Flood season 0.47 Low-flow season 0.004
Low-water-level operation	Season		Low-flow season May Flood season June-October	Low-flow season March-May
	Average discharge	(m^3/s)	18,000	540
	Sediment inflow	(100 million tonnes)	444	0.048

Note: Data for Danjiangkou Reservoir are from Baihe Hydrological Station, Hanjiang River.

maintain the natural conditions of the river channel. To improve the shipping conditions of the port and the reservoir area, it is necessary to control the minimum water level in the dry season.

At present, the depth of the navigable channel maintained at Chaotianmen dock is 2.9 m. The equivalent elevation of the water level is 160 m. If the condition of ensuring a direct river link to Chongqing for ships in the 10,000-t class is to be met, the depth of the navigable channel has to be increased by 3.5 to 4.0

m, and the elevation of the water level at Chaotianmen dock cannot be lower than 161 m.

This is the water level that will have to be maintained during the falling stage of the dry season. A scheme we proposed in August 1983, in which a normal pool level of 170 m, a minimum water level of 160 m in the dry seasons, and a limiting water level of 145 m in the flood season were stipulated (or the 170–160–145-m scheme), was designed to ensure a direct river link to Chongqing by ships in the 10,000-t class. When the water level at the dam is 160 m, the water level at Chaotianmen dock will exceed 161 m because of the backwater effect. According to this scheme, the reservoir level will not exceed 145 m in the flood season, which will likely prevent the expansion of the deposit to the port of Chongqing and thus the adverse effects of sedimentation. At the same time, sufficient water depth is maintained at the dam during the dry season to ensure adequate depth of the navigation channels at the port area of Chongqing and in the lower reaches. This is our basic idea for solving the navigation problems in the FBZ of the TGR by adopting a rational water level scheme and regulating method.

It should be pointed out, however, that the minimum water level in the dry season is related to the regulating pool. The higher the minimum water level, the smaller the regulating pool and the lower the guaranteed output of hydropower and the navigable flow downstream. Therefore it is necessary to make a comprehensive analysis and comparison of the data concerned before the rational minimum water level in the dry season can be selected.

It should also be noted that the water level in the dry season does not necessarily fall to the lowest level every year. In fact, only in the design dry year will the water level decline to its minimum toward the end of the dry season. The year 1960 is the design dry year for the TGR, with an exceedence probability of 0.97. Therefore, for most of the other years, the water inflow will be larger than that in 1960, and the water level by the end of the dry season will also be higher than the minimum water level.

In the 170–150–140-m scheme, for instance, although the regulated water level during the falling stage in the dry season is 150 m, in fact there are only twenty-five out of one hundred years when the reservoir level will decline to 160 m or lower, and on average there will only be twelve days per year in the dry season when the water level is lower than 160 m [10]. Thus, the selection of the minimum water level must be considered in conjunction with the exceedence probability for the depth of navigable channels. After the exceedence probability has been rationally determined, the designed minimum water level can be appropriately lowered in order to increase the output of hydropower and the navigable discharge in the lower reaches.

Inadequate navigable depth may pose problems for navigation in the middle and lower sections of the FBZ as well as the upper end of the perennial backwater zone after construction of the TGR and as a result of processes of scour and fill, changes in the river regime, and migration of the navigable channel. Besides local treatment, appropriate changes in the regulating process can also be employed to solve these problems. Because the TGR has a relatively small storage capacity, control and regulation of its water level is relatively easy. Such methods should be seriously considered for adoption.

3. Will construction of the TGP transfer the flood prevention problem from the lower to the upper reaches of the Changjiang River?

After the construction of the TGP, the area inundated will be expanded. This is because of the elevation of the river bed due to sedimentation and the backwater effect of the dam. The flood stage will also be raised. Therefore, predicting the flood heights after construction of the TGR has become a popular concern. Some suggest that the hundred-year flood stage will be raised to greater than 210 m at Chongqing, and to greater than 220 m at Jiangjin and Hechuan, implying that the backwater will enter the Sichuan Basin. Thus, they believe that the construction of the

TGP will transfer flood disasters from the lower to the upper reaches of the Changjiang River, and therefore they conclude that the TGP should not proceed in the short run [11]. Others argue, however, that according to the mode of operation specified in the design of the TGP, the flood stage at Chongqing will not be raised very much. Such changes in flood stage will only have limited impact and can easily be controlled. These two opinions are diametrically opposed, and their validity will certainly have a great influence on the decision to approve the project designs. In view of this situation, we believe it is essential to discuss and analyze this problem to arrive at a more realistic assessment of the situation which will serve as the basis for decision making.

3.1. The magnitude of the raised flood stage after construction will be limited and can be controlled

When a reservoir is built, the flood stage in the reservoir area depends not only on the water level at the dam, but also on the shape and size of the sediment body. The former is determined by the limiting water level in the flood season, and the latter is reflected mainly by the equilibrium gradient. In the following, we will discuss these problems with reference to the conditions of the TGR.

According to the theory of river channel metamorphosis, following the construction of a reservoir, the river channels in the reservoir area will experience a period of readjustment. The changes to be expected are characterized by the elevation of the river bed due to sedimentation, reduction in size of the bed material, narrowing and deepening of the channel cross-section, and lowering of the channel gradient. Such changes will continue until the channel has the capacity to transport all the water and sediment derived from upstream. In general, the new equilibrium gradient (J) is lower than the original gradient (J_O). However, the ratio J/J_O depends on the physical conditions of the river. Because of their steep gradient, the bed material of mountain streams is usually coarser and the stream is often undersaturated with sediment. Because of the reduced intensity of flow after construction of the reservoir, part of the wash load derived from the drainage basin will be deposited on the river bed. Thus, the bed material will become much finer, and this will reduce bed resistance to flow and increase the supply of fine material. As a result, a lower equilibrium gradient is required to transport the same amount of sediment. In the case of reservoirs built on plains, the original bed material is quite fine, and little change in bed material size is expected after construction. Thus, the new equilibrium gradient will approximate the original gradient. In figure 6, data derived from twenty-two reservoirs in China [12] are plotted to show the relation between D/D_O and J/J_O where D and D_O are the median size of bed material before and after reaching the new equilibrium condition. Although the scatter is quite large, it is obvious that the trend is for the new equilibrium gradient to decrease when D/D_O increases.

The river channels in the TGR area have characteristics of mountain streams. Under natural conditions, the bed material is mostly gravel and bedrock, with a channel gradient of 0.002. The equivalent grain size of bed material is 10–90 mm, with a maximum of 150–200 mm [13]. The stream is usually undersaturated with sediment, and most of the suspended sediment is wash load. After the completion of the TGR, deposits will be mostly medium and fine sand derived from upstream. Bed material will have a median diameter of 0.15–0.20 mm, and a D/D_O ratio of less than 0.02. According to figure 6, the new equilibrium gradient will be 10 to 35 percent of the original, suggesting that the future equilibrium gradient of the TGR will be relatively gentle.

Computations based on mathematical models developed by the Academy for Changjiang River Studies and the Chinese Academy of Hydraulics suggest that when the TGR reaches the equilibrium state in sedimentation, the channel gradient will be about 0.00007, which is 35 percent of the original. For the purpose of assessing the rationality and reliability of this result, we have developed the following equation for the equilibrium gradi-

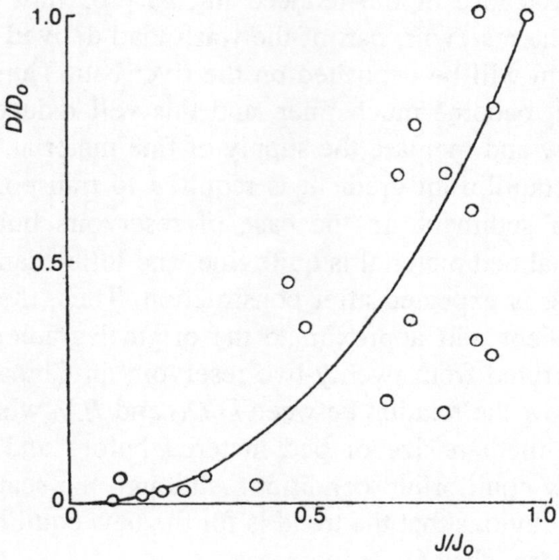

Figure 6. Relation between D/D_o and J/J_o

ent of sand bed streams by using the concept of sediment transport equilibrium and actual data collected from rivers and associated reservoirs in the middle and lower reaches of the Changjiang River, Huanghe River, and Hanjiang River.

$$J = 0.0062\, D_{50}{}^{1.4}/(Q/B)^{0.92} \ldots (1)$$

where J is the equilibrium gradient, D_{50} median bed material size [mm] after reaching the equilibrium in sedimentation, Q mean discharge [m^3/s], and B river width [m]. Figure 7 shows the goodness of fit of this estimation formula to the measured data [14]. Where median bed material size, mean discharge, and river width are known, equilibrium channel gradient can be estimated by using equation 1 (table 5). It can also be seen that using 0.007 as the estimated equilibrium gradient has a considerable degree of reliability.

Located immediately downstream of the Three Gorges, the

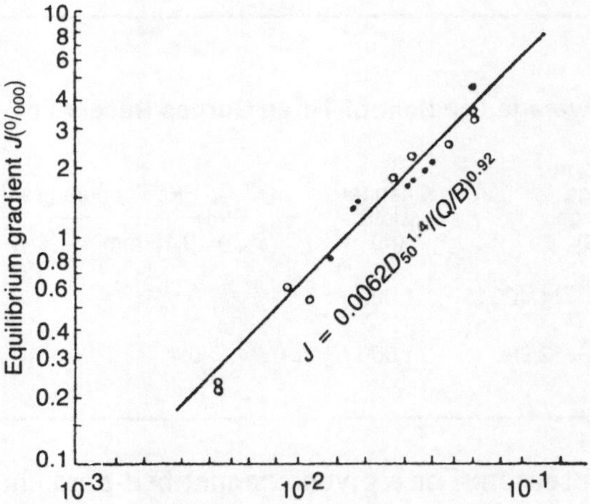

Figure 7. Relation Between J and $D_{50}^{1.4}/(Q/B)^{0.92}$

Upper Jingjiang section, which forms part of the Jianghan alluvial plain, has similar sediment and water inflow conditions as the reservoir area. The conditions for sedimentation and channel bed metamorphosis are also similar. Therefore, the channel gradient of the upper Jingjiang section serves as an important reference for the reservoir area. The average gradient of the 200-km-long river course of the upper Jingjiang section (from Yichang to Shishou) is 0.00005. This is generally the same as the surface gradient of the deposits in the reservoir. Gradient of the upper end of the deposit, which is taken to be 0.000125, is approximated by the average of the natural channel and the surface gradient of the deposit. Thus, the weighted mean gradient will not exceed 0.00007.

It can be concluded from the above discussion that the equilibrium gradient estimated by using mathematical models is basically reliable and can be used as the basis for estimating future flood stages at Chongqing.

River sediment load is usually concentrated in the flood season, and thus processes related to river bed changes are the most intensive in this period. In other words, the amount of deposition in the reservoir area will depend on the hydrological conditions

Table 5

Estimated Average Gradient of Three Gorges Reservoir

	Long-term average discharge (m^3/s)	Channel width (m)	Equilibrium gradient	
			D_{50} = 0.15 mm	D_{50} = 0.20 mm
Cuntan	11,400	1,000	0.5	0.7
Cuntan + Wulong	12,990	1,000	0.4	0.6

in the flood season. For a given channel bed gradient, the lower the limiting water level, the smaller the amount of reservoir sedimentation, and the less the expansion of accumulated sediment upstream. Thus, the flood stage at the upstream end of the reservoir area is less affected. The reverse is true for a high limiting water level. Therefore, by adjusting the limiting water level in the flood season, the flood level at the upper reaches of the reservoir region can be fundamentally controlled.

The Academy for Changjiang River Studies has attempted to predict the flood stages in Chongqing after the construction of the TGR by using mathematical models. The results suggest that the flood stages at Chongqing will depend mainly on the limiting water level in the flood season and on the number of years the reservoir is in operation. The relations among these variables are shown in figure 8. It can be seen that the limiting water level and the flood stage exhibit a direct linear relation. If the limiting water level is raised by 10 m, this will cause a rise of the flood stage by 2.5–3.0 m. The influence of the duration of reservoir operation on the flood stage will diminish with time. Generally, after eighty years of operation, sedimentation will be stabilized, and so will be the flood stage [15].

Among the major alternatives for the TGP at the present time is that the design limiting water level in the flood season varies between 135 m and 150 m. If a flood with a discharge of 82,800

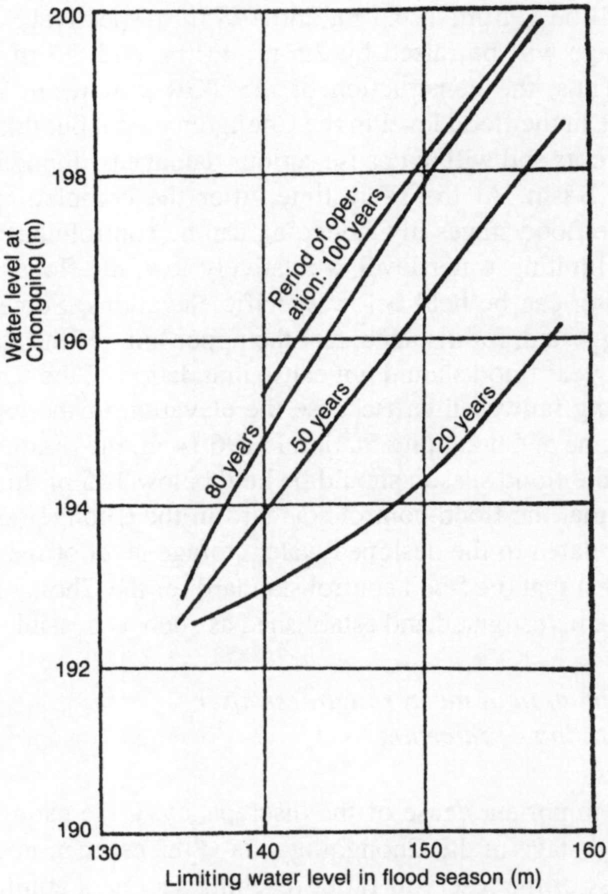

Figure 8. Relation between flood level at Chongqing and the limiting water level in flood season (Flood discharge: 82,800 m^3/s)

m^3/s occurs after the reservoir has been operating for eighty years (the flood discharge in 1981 was 82,800 m^3/s, roughly equivalent to a hundred-year flood; in the 800-year period since A.D. 1153, eight floods have had a discharge greater than or equal to 80,000 m^3/s), and if the limiting water level is 135 m, the flood stage in Chongqing will be similar to the natural flood level, and virtually no change can be expected. If the limiting water level in the flood season is 140 m, 145 m, and 150 m, then the corresponding flood level will be 195 m, 196.5 m, and 198 m, respectively, and the flood stage will be raised by 2.5 m, 4.0 m, and 5.5 m, respectively. Thus, the construction of the TGR will result in some increases in the flood level in the Chongqing area, but this will be within limits and will not cause serious damage by inundating the Sichuan Basin. At the same time, after the completion of the TGR, the flood stages in Chongqing can be controlled. Provided that the limiting water level is relatively low, the flood stage in Chongqing can be held below specific elevations. Some people have suggested, for instance, that the upper limit of the stage of a hundred-year flood should not cause inundation of the Chengdu–Chongqing railway line. Because the elevation of the top of the railway line at Chongqing Station is 196.14 m, the limiting water level in the flood season should be held below 145 m. In view of the fact that the flood control standard in the Chongqing area is closely related to the designed water storage level of the project, we suggest that the flood control standard for the Chongqing area should be investigated and established as soon as possible.

3.2. Estimation of mean roughness after sedimentation equilibrium

Another important cause of the discrepancy in the estimated future flood stage in the Chongqing area is the estimation of mean roughness of the reservoir (after reaching sediment equilibrium). One author has pointed out that, after the TGR is stabilized, if a flood with a discharge of 66,500 m^3/s occurs, the flood stage at Chongqing will reach 209 m, more than 20 m higher than that estimated by using mathematical models [16]. It was suggested that the estimates calculated on the basis of mathematical models should be reevaluated. We have scrutinized the different results from the two sources and found that the above-mentioned author did not take into account the backwater effect of the reservoir. This resulted in an underestimation of the area of the channel cross-section and an overestimation of flow velocity and flood stage. Another important reason is that different mean roughness coefficients have been employed. Table 6 is a comparison of the mean roughness coefficients employed in these sources.

The mean roughness of the reservoir has two components, bed roughness and bank roughness. After the TGR reaches an equilibrium state in sedimentation, bedrock and gravel on the channel bed will be buried by thick layers of sandy deposits, and bed roughness will be greatly reduced. Based on data collected in the upper Jingjiang section, after the reservoir reaches an equilibrium in sedimentation, the median size of the bed material will be 0.15–0.20 mm. When water discharge exceeds 60,000 m^3/s, the water depth in the reservoir area will be 30–40 m [17]. Under these conditions, the bed roughness coefficient will be in the range of 0.017–0.020. A coefficient of 0.033 was used in one of the calculations [16], which represents that of a bed of gravel [13]. This value is obviously an overestimation.

The TGR is a canyon-type reservoir. After its completion, there will be increases in the water level and water depth. The significance of bank roughness in relation to mean roughness will increase. Therefore, attention should be paid to evaluating the reliability of the bank roughness coefficient. Based on measured data, we have estimated the bank roughness of every river section in the reservoir area by using a variety of methods [18]. The fact that these coefficients are consistent with those employed in the mathematical models suggests that the latter are reliable.

If the above-mentioned bed roughness and bank roughness coefficients, which we think are reasonable, are used to estimate the mean coefficient, we will find that the mean roughness coefficient employed in the mathematical models are also reasonable, and those used in reference 16 are overestimated.

Relative to the size of the basin in the lower reaches of the Changjiang River and its huge flood volume, the flood control capacity of the TGR is limited in effect. However, the flood control function of the TGP is obvious for the Jingjiang section, which has the most serious flood hazard. At present, without resorting to the flood diversion area, the Jingjiang embankments can only protect the area from floods with a return period of ten

Table 6

Comparison of Roughness Coefficient Employed in the Mathematical Model and in Reference 16

River reach	Mathematical model	Reference 16
Dam site-Fengjie	0.036	0.045
Fengjie-Zhongxian	0.022—0.025	0.036
Zhongxian-Cuntan	0.020-0.028	0.034

to twenty years. After the construction of the TGR, it can be protected from hundred-year floods. To maintain the storage capacity and control the flood stage upstream, the TGR will have to be operated at a low water level in the flood season. It will serve the purposes of power generation, navigation, and, above all, flood control. These are precisely the comprehensive benefits brought about by the TGP. Hence, while it is improper to overstate the flood control benefit of the TGP, to neglect the flood control benefits or even to suggest that the project will aggravate the flood hazard of the area is both unfair and unfounded.

4. Will the construction of the TGP cause substantial coastal retreat and severe salinity incursion at Changjiang River estuary?

4.1. Why is construction of some reservoirs significant for the estuary area?

The estuary is the outlet of the river where it enters the sea. Here, it is simultaneously affected by the dynamics of the riverine and marine processes, producing a specific type of landform, ecological environment, and pattern of channel metamorphosis different from that of the river proper. Due to the interactive effects of the riverine and marine processes, a variety of estuary types can be produced. When a reservoir is built, the amount and process of

Table 7

Comparison of Bank Roughness Coefficient Employed in the Mathematical Model and in Reference 18

River reach	Bank roughness coefficient	
	Mathematical model	Reference 18
Wanxian-Yunyang	0.0394	0.0560
Yunyang-Fengjie	0.0523	0.0510
Fengjie-Wushan	0.1249	0.1110
Wushan-Badong	0.0804	0.1030
Badong-Zigui	0.0965	0.1060
Zigui-Taipingxi	0.1093	0.0920

water and sediment inflow in the estuary area will be altered to some extent. In turn, this will disrupt the equilibrium condition in the estuary area, which has developed over a long period of time, causing various adverse impacts. It is evident that the larger the reservoir's regulating capacity, the greater will be the changes in water and sediment transport, and, therefore, the greater and more profound will be the impact on the estuary area. Take the Aral Sea in the Soviet Union as an example. Because the runoff of the Amu-Dar'Ya River and Syrdar'Ya River, which are the sources of water supply for the Aral Sea, has been detained and used for irrigation, the sea level has been lowered by 3 m, and the coastline retreated by 50 km during 1966–1976, causing heavy losses in the fishery industry and severe salinization in the surrounding area. Another example is the Volta River in Ghana. During the water-impounding period following completion of a reservoir in the river, the problem of salinity incursion was aggravated. The incursion distance of the 100 ppm salinity isopleth increased from 30 km to 50 km.

The problems caused by the construction of the Aswan High Dam in Egypt are well-known. The sediment-trapping effect of the reservoir on the Nile deprives the estuary area of sediment supply, causing coastal erosion, and therefore has an adverse impact on the tourist industry. Furthermore, the reduction of fresh water input into the estuary led to the migration of the 39-percent-salinity isopleth from 80 km off the coast to the vicinity of the coastline. The retention of river water and reduction in peak flow by the reservoir caused a dramatic decrease in the sardine catch in the Mediterranean Sea and eliminated the input of nutrients by flood water in the delta area. All the above examples suggest that reservoir construction has a significant impact on the estuary area because these reservoirs have a large storage capacity and significant regulating function, which have led to important changes in the hydrological and sedimentary processes at the estuary. The example of the reservoir at the Aswan High Dam will be discussed in detail.

The Aswan Reservoir was built on the upper reaches of the Nile, more than 900 km upstream of the estuary. It has a total storage capacity of 164 billion m^3, of which 39 billion m^3 is the designed sediment pool and 125 billion m^3 the effective storage capacity. The long-term (1871–1962) average annual runoff that enters the sea is 33 billion m^3. Thus, the effective storage capacity is 3.78 times that of the river's annual runoff. This suggests that the Aswan Reservoir has a large regulating capacity, and that the reservoir can be used to regulate multiyear river flow. After the construction of the reservoir, downstream discharge was fundamentally changed. Take flood characteristics as an example. Before construction, almost all the runoff of the Nile from August to November was discharged into the Mediterranean Sea. The fresh-water input has an important influence on the salinity distribution of the estuary area. It lowered the salinity of the surface water in the offshore area to 28 percent or less. After construction, most of the flood water was regulated and used for irrigation. This, in conjunction with the huge evaporation losses, has resulted in a negligible reservoir outflow. Figure 9 shows the changes in measured monthly water discharge at the estuary of the Nile. It can be seen that after the completion of the reservoir in 1964, the discharge at the estuary was greatly reduced, with the peak flow reduced to only one-seventh of the original. As a result of these changes, the 39-percent-salinity isopleth, which represents the average condition of sea water in the Mediterra-

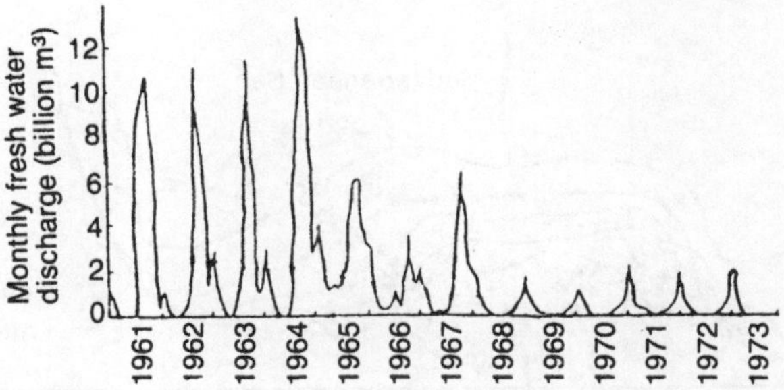

Figure 9. Variations in fresh water discharge at the Nile estuary before and after construction of the reservoir

nean, was shifted and was encroaching upon almost the entire coastline (figure 10) [19–23].

Before the completion of the Aswan Reservoir, the annual sediment load of the Nile varied from 60 to 180 million tonnes. Much of the fine-grained sediment was carried to the deep sea, and the coarser sediment was deposited in the estuary, forming the fertile Nile Delta. Over a long period of time, an equilibrium between sediment supplied by the river and coastal erosion by the sea currents was established, and specific coastal landforms were developed. After the completion of the reservoir, almost all the sediment was trapped by the reservoir. Outflow from the reservoir is minimal and can only scour a very limited amount of sediment from the river bed. Table 8 shows the annual sediment load at the Aswan Dam site and at the estuary of the Nile from 1964 to 1982. In 1964, the annual sediment load discharged into the sea was more than 50 million tonnes. In the 1980s, the figure declined to 2–3 million tonnes, or only 5–6 percent of the original load. Under these circumstances, the sediment balance at the estuary was completely disrupted, and coastline retreat as a result of erosion is to be expected [19–20].

4.2. The TGR should have no major impact on the estuary area

Considering the tremendous environmental and ecological impact of the Aswan Reservoir on the estuary area of the Nile, some

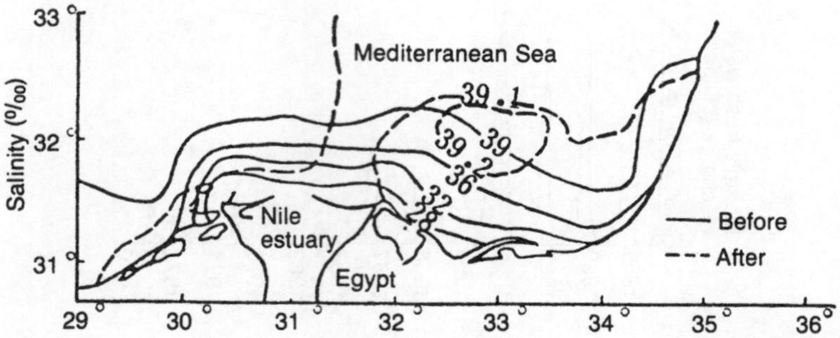

Figure 10. Salinity ($^0/_{00}$) distribution at the Nile estuary before and after the construction of the Aswan Reservoir

people have expressed great concern whether the construction of the TGR will bring a variety of adverse impacts on the prosperous estuary area of the Changjiang River. We believe that the TGR exhibits different characteristics from the Aswan Reservoir, and its impact on the estuary area will be totally different as well.

The total storage capacity of the TGR is 34.5 billion m^3 (in the 170–160–145-m scheme), of which about 17 billion m^3 is for sediment storage, and thus 17.5 billion m^3 is the effective capacity. When compared to the long-term mean annual runoff volume of 450 billion m^3 at the Yichang hydrological station and 908 billion m^3 at the Datong hydrological station, the capacity-inflow ratio is only 0.04 and 0.02 respectively. The TGR, therefore, is a reservoir that has a small regulating capacity and can only regulate the seasonal flow. Thus, it is expected that there will be only a few months of the natural pattern of flow. According to the present designed mode of operation, gates in the dam will be kept open in the flood season, and the natural changes in flow rates will basically be unaltered. By the end of the flood season, the reservoir will begin to impound water and will reach its normal level of 170 m in October-November, storing a total volume of 17.5 billion m^3. During January-April, 8.2 billion m^3 of water will be released. By the end of May or the beginning of June, the reservoir level will decline from 160 m to 145 m. The mean

Table 8

Sediment Load at Aswan Dam Site and Nile Estuary

Year	Sediment load at dam site (10^6 t)	Sediment load at estuary (10^6 t)
1964	26.27	51.76
1966	3.78	9.04
1968	2.30	3.22
1970	2.67	4.44
1972	2.71	4.40
1974	2.83	5.08
1976	1.61	3.90
1978	2.20	4.70
1980	1.79	3.32
1982	1.92	2.60

monthly discharge under natural and regulated conditions at the Datong hydrological station is listed in table 9. It can be seen that with regulation, the river flow increases slightly in the dry season but decreases slightly by the end of the flood season. Discharge remains the same during the flood season and in December. Changes in the total annual flow are very small.

What will be the effect of the above-mentioned changes in river discharge on salinity incursion in the estuary area of the Changjiang River? Reference 24, which is a study of the impact of the south-north water transfer project on salinity incursion in the same estuary, indicates that the degree of salinity incursion depends mainly on the river discharge. The greater the discharge, the smaller is the degree of salinity incursion, and vice versa. In Shanghai, water for industrial, agricultural, and domestic use is mainly derived from the Huangpu River. The waterworks affected by salinity incursion are located at Wusong, Zhamen, and Yangshupu. The Wusong waterworks is located closest to the estuary, so its water quality may be used as a criterion for judging the effect of salinity incursion. Chlorine content is usually used to measure water quality. A chlorine content of 100 ppm indicates that the water quality has been affected by salinity in-

Table 9

Comparison of Natural and Regulated Discharge of Three Gorges Reservoir in a Year of Average Flow at Datong Hydrological Station

Month	Natural discharge (m^3/s)	Regulated discharge (m^3/s)	Difference (%)
January	10,200	11,170	+9.5
February	11,200	12,470	+11.3
March	15,200	15,970	+4.6
April	23,300	23,440	+0.6
May	35,500	39,100	+11.0
June	41,100	41,100	0
July	48,900	48,900	0
August	45,600	45,600	0
September	42,000	42,000	0
October	35,800	33,200	+7.3
November	25,200	21,200	−15.9
December	14,900	14,900	0

cursion, and this will have a severe impact on factories requiring high-quality water. People can also sense this by drinking water. When the chlorine content reaches 500 ppm, agricultural and domestic use of the water will be seriously affected. Based on data collected in the estuary of the Changjiang River, it was suggested [24] that there is a significant relationship between the number of days per year when discharge at the Datong hydrological station is less than 14,000 m^3/s and the number of hours per year when the chlorine content exceeds 100 ppm or 500 ppm at the Wusong waterworks. Figure 11 shows these relations, which can be expressed in the following equations:

$$T > 100 = 158\, e^{0.0223N} \quad \ldots (2)$$
$$T > 500 = 9.4\, e^{0.0382N} \quad \ldots (3)$$

where T is the number of hours per year when chlorine content exceeds 100 ppm or 500 ppm at the Wusong waterworks, and N, number of days per year when discharge is less than 14,000 m^3/s at the Datong hydrological station. Based on these two equations,

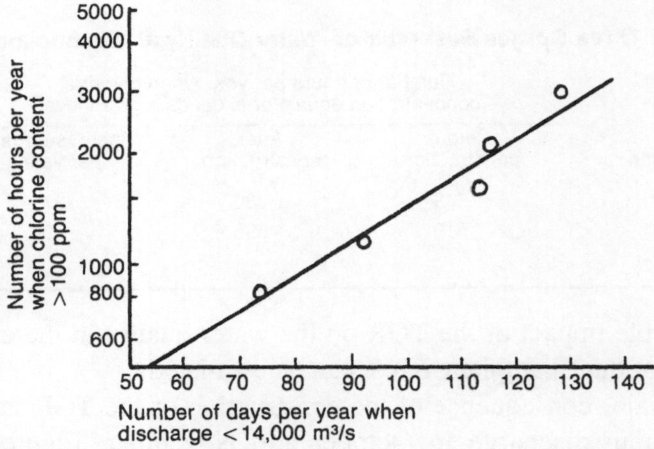

Figure 11(a). Relation between chlorine content (>100 ppm) and discharge at Wusongkou

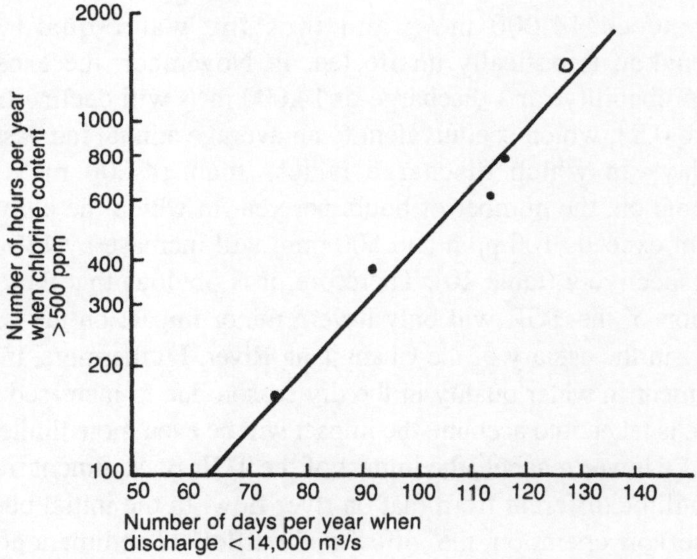

Figure 11(b). Relation between chlorine content (>500 ppm) and discharge at Wusongkou

Table 10

Impact of Three Gorges Reservoir on Water Quality at Wusongkou

Chlorine concentration	Number of hours per year when chlorine concentration equals or exceeds a given level: Before construction	After construction	Increase (hrs) per year
≥100 PPM	1,643	1,780	137
≥500 PPM	519	595	76

the possible impact of the TGR on the water quality in the estuary area of the Changjiang River can be estimated.

The major consequence of the construction of the TGR is the reduction in discharge in October and November. Figure 12 illustrates the exceedence probability of discharge at the Datong hydrologic station before and after the construction of the reservoir. It shows that after construction, discharge in October will still exceed 14,000 m^3/s, and thus the water quality at Wusongkou is basically unaffected. In November, the exceedence probability for a discharge of 14,000 m^3/s will decline from 0.95 to 0.83, which is equivalent to an average annual increase of 3.6 days in which discharge is less than 14,000 m^3/s. At Wusongkou, the number of hours per year in which the chlorine content exceeds 100 ppm and 500 ppm will increase by 137 and 76, respectively (table 10). Therefore, it is obvious that the construction of the TGR will only have a minor impact on the water quality in the estuary of the Changjiang River. Furthermore, if improvement in water quality in the dry season due to increased discharge is taken into account, the impact will be even more limited.

In the lower reaches, the impact of the TGR on sediment transport will be different from that on river flow. In the initial period of reservoir operation, the inflow and outflow of sediment at the reservoir will not be balanced. About 60 percent of the sediment will be trapped in the reservoir, and the other 40 percent will be discharged. For a discharge of less than 60,000 m^3/s, the TGR will be operated in such a way to allow the water to flow through.

During major floods, the reservoir will only marginally reduce the flood peaks, and thus there will be little change in the carrying capacity of the river. Therefore, when the river flow that is below sediment capacity is transmitted to the lower reaches, it will cause some scouring along the way, but by the time it reaches the estuary, after a journey of more than 1,000 km, it will be loaded with sediment and will have a sediment concentration similar to that under natural conditions. The experience at the Sanmenxia Reservoir can be used to illustrate this point. From September 1960 to October 1964, the total amount of sediment input and output was 6.76 billion t and 2.17 billion t respectively. In addition, the sediment supply was 2.31 billion t in the lower reaches. Thus, 4.48 billion t of sediment were transported to the estuary, which is equivalent to 66 percent of the sediment inflow. Under natural conditions, the sediment delivery ratio of the river channel in the lower reaches of the Huanghe River is 75 percent. In other words, without the Sanmenxia Reservoir, the amount of sediment discharged into the sea during those four years would have been 5.07 billion t, only 13 percent more than the case with the reservoir. This shows that provided that discharge is not reduced, and that flood peaks are not lowered too much, sediment load transported to the estuary will be minimally affected during the early period of the TGR's operation. After fifty years of operation, the sediment delivery ratio at the reservoir will increase to 80 percent, and after eighty years it will be higher than 90 percent. By that time, most of the inflow sediment will be discharged, a situation that is virtually the same as that before construction.

In summary, the TGR has a small regulating capacity and will only affect the pattern of seasonal discharge. It will have limited impact on water discharge and sediment load in the middle and lower reaches of the Changjiang River and its estuary. Therefore, no major impact on the geomorphology, environment, and ecology in the estuary area is expected. The TGR has its own unique characteristics, which are entirely different from those of the Aswan Reservoir. Therefore, it is not advisable to make simple inferences of the conditions of the TGR after its completion from known problems of the Aswan Reservoir.

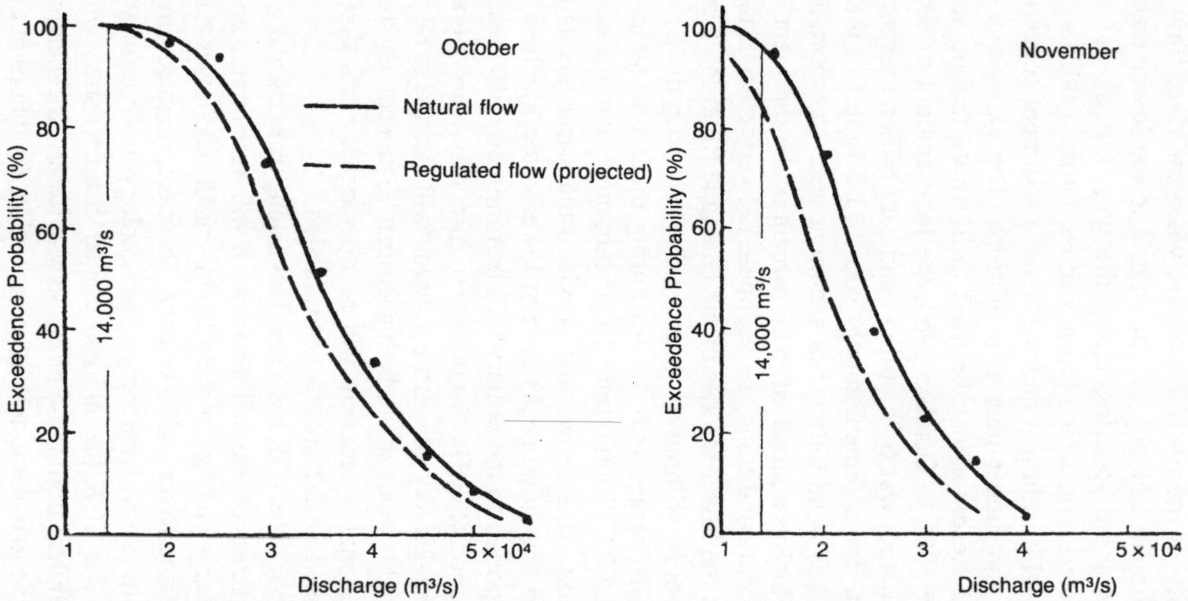

Figure 12. Exceedence probability of water discharge at the Datong Hydrological Station, Changjiang River (1922–1972)

Literature cited

[1] Yellow River Committee, Ministry of Water Conservancy and Electricity of China, "Huanghe liuyu Sanmenxia shuiku kuqu shuiwen shiyan ziliao" (Hydrological experiment data in the Sanmenxia Reservoir region of the Huanghe River valley), 1960–1980.

[2] Office of the Changjiang River, "Changjiang Sanxia shuili shuniu zhengchang xushuiwei buchong lunzheng baogao" (A supplementary report on normal storage level of the Three Gorges Reservoir), 1985.

[3] Hydrology Bureau, Office of the Changjiang River, "Changjiang zhongxiayou hedao jiben tezheng" (The basic characteristics of river channel in the middle and lower reaches of the Changjiang River), Wuhan, 1983.

[4] Shanxi Provincial Institute of Water Conservancy, Sediment Research Laboratory of Qinghua University, *Shuiku nisha* (Reservoir sedimentation) (Beijing: Shuili dianli chubanshe, 1979).

[5] Bureau of Water Conservancy, Ministry of Water Conservancy and Electricity of China, "Quanguo zhuyao heliu shiuwen tezheng tongji" (Hydrological characteristics of the major rivers in China), 1975.

[6] Sediment Research Laboratory, Department of Water Conservancy, Qinghua University, "Changjiang Sanxia shuili shuniu 150 fangan yunyong zhong hou qi Lanzhuba heduan nisha moxing shiyan baogao" (The modeling experiment for the 150-m dam scheme on Lanzhuba river reach's sedimentation during the middle and later period of operation of the Three Gorges Reservoir), 1985.

[7] Jiao Enze and Lin Binwen, "Xiaolangdi shuiku keyong kurong ji chongyu xingtai gusuan" (The effective capacity of the Xiaolangdi Reservoir and the estimation of its alluvium), Institute of Water Conservancy, Huanghe River Committee, 1979.

[8] Danjiang Hydrological Station, "1960–1974, 1975–1980 nian Danjiangkou shuiku shuiwen nisha shiyan ziliao" (Hydrology and sediment experimental data of the Danjiangkou Reservoir, 1960–1974, 1975–1980), 1975, 1981.

[9] ______, "Guanyu Changjiang Sanxia shuniu zhengchang xushuiwei he yunyong fangshi de lunzheng" (On the normal pool level and operation mode of the Three Gorges Reservoir), 1985 (third draft).

[10] ______, "Butong xushuiwei fangan dui tonghang shijian de yingxiang" (The impact of different pool levels on navigation), 1985.

[11] Economic construction group, Chinese People's Political Consultative Committee, "Sanxia gongcheng jinqi buneng shang" (The Three Gorges Project should not go ahead in the short term), 1985.

[12] Zhang Ren, Qian Ning, "Reservoir Sedimentation," IRTCES, Sediment Research Laboratory, Qinghua University, Beijing, 1985.

[13] Hui Yujia and Chen Zhicong, "Changjiang Sanxia hedao zaolu de chubu fenxi" (A pıeliminary analysis of the Three Gorges' channel roughness),

Shuili xuebao (Bulletin of water conservancy), 1982.

[14] Wang Shiqiang, "Changjiang Sanxia shuiku pingheng zongpoumian he yuji moduan yugu" (Longitudinal profile of the Three Gorges Reservoir and estimation of the sediment wedge) (Beijing: Qinghua Daxue shuili xi nisha yanjiushi, 1982).

[15] Changjiang Academy of Water Conservancy and Hydro-Electricity, "Sanxia shuili shuniu nisha yuji jisuan chengguo ji fenxi" (The calculation result and its analysis for sediment of the Three Gorges Water Conservancy Project—a progress report), 1986.

[16] Fang Zongdai, "Sanxia huishuixian jianhuafa ji jisuan chengguo" (A simplified method of determining the backwater line of the Three Gorges Reservoir and its calculation results) (Beijing: Institute for Sediment Studies, Chinese Academy of Water Conservancy and Hydro-Electricity, 1986).

[17] Han Qiwei, He Mingmin, and Sun Weidong, "Sanxia shuiku 150 fangan xuanyizhi yuji de jisuan yu fenxi" (Calculation and analysis of suspended load sedimentation for the 150-m dam scheme of the Three Gorges Reservoir) (Beijing: Chinese Academy of Water Conservancy and Hydro-Electricity, 1984).

[18] Wang Shiqiang, Song Genpei, and Bi Changshou, "Changjiang Sanxia shuiku xushui chu zaolu fenxi" (Analysis of the roughness of the Three Gorges Reservoir during the initial period of water storage) (Beijing: Qinghua Daxue shuili xi nisha yanjiushi, 1981).

[19] Petts, G. E., *Impounded River* (Loughborough, Leicestershire, England: University of Technology, 1984).

[20] Din, S. H. Sharaf, "Effects of the Aswan Dam on the Nile Flood and on the Estuarine and Coastal Circulation Pattern Along the Mediterranean Egyptian Coast" *Limnology and Oceanography* 22, 2:194–207.

[21] Kinawy, I. E., et al., "Some Effects of the High Dam on the Environment," in *Transactions of the Eleventh International Congress of Large Dams*, Madrid, 1973, 1:959–73.

[22] Buttling, S., et al., "Predicting the Rate and Pattern of Storage Loss in Reservoirs," in *Transactions of the Eleventh International Congress of Large Dams*, Madrid, 1973, 1:565–80.

[23] Shalash, S., "Degradation of the River Nile," *Water Power and Dam Construction* (July 1983, August 1983).

[24] Han Naibin et al., "Nanshui beidiao dui Changjiang kou yanshui ruqin de yingxiang" (The impact of the South-North Water Transfer Project on the intrusion of salt in the Changjiang River's estuary area) (Nanjing: Nanjing Institute of Water Conservancy, 1978).

[25] Qian Ning, Zhang Ren, and Zhou Zhide, *Hechuang yanbianxue* (River bed evolution) (forthcoming from Kexue chubanshe).

[26] Cai Tingsheng et al., "Guowai shuili shuniu huishui biandong qu de yuji jiqi dui hangyun de yingxiang" (The sedimentation in variable backwater areas and its impact on navigation in foreign countries) (Nanjing: Nanjing shuili kexue yanjiuyuan, 1985).

6

The Flood Prevention Function of the Three Gorges Project—Disadvantages Outweigh Advantages

Fang Zongdai

THE THREE GORGES PROJECT (TGP) is a gigantic water conservancy project that has attracted the attention of the world and the concern of the whole country. The project is designed to prevent flooding, improve navigation, and generate hydropower. But flood prevention is considered the most important, as other projects may be built to generate hydropower, and navigational conditions at the upper end of the reservoir will, in any case, deteriorate after the completion of the dam. Our analysis of the flood prevention function of the TGP shows that more harm than benefit will be created. Whether the TGP should be constructed is a major issue that deserves detailed reassessment.

Each time floods occur on the Changjiang (Yangtze) River, the amount of flood water is enormous. Using the 1954 flood as an example, the total amount of "excess flood water" (i.e. flood water requiring regulation) was 102.3 billion m^3. Since then, the embankments have been raised, reducing the total amount of excess flood water to 50 billion m^3. The plans of the TGP call for

---

Fang Zongdai, "Sanxia gongcheng fanghong xiaoyi bi li shao." *Keji daobao* (Science and technology review) 4 (1986): 17, 31–34. The author is a senior engineer with the Water Conservancy and Hydropower Research Institute. Translated by Jane Tan.

a flood prevention capacity of 22 billion m^3 above the regular levels needed for hydropower generation. This amount is less than half of the excess flood water requiring regulation. Therefore, the TGP has a limited capability to prevent flooding. More importantly, however, the construction of the dam will bring about changes in the hydrological and sedimentary characteristics of the river both upstream and downstream and adversely affect the flood prevention capability of the cities of Chongqing, Hankou, and Shanghai. This point was raised at the July 1985 meeting to discuss the sedimentation and navigation problems associated with the TGP. Until now, however, it has not received the attention of the departments concerned, and this is why it is reiterated in detail here.

1. Chongqing will be inundated by artificial floods because the designed flood stage is significantly higher than the historical maximum level.

At the macro level, the TGP will create a typical canyon-type reservoir, which can be divided into three sections according to the size of the valley. The narrowest section, from Fengjie downstream to the dam at Sandouping, is about 200 km long and contains the world-famous tourist sites. Along this section of the canyon are the Qutang, Wushan, and Xiling gorges. The minimum width is between 200 and 300 m, and it varies from section to section. The flow is rapid, and resistance to flow is high. Because of bank roughness and channel loss, water surface gradient is steep. According to actual data collected, when water discharge Q = 110,000 m^3/s (which was recorded in 1870), the water surface gradient S would reach 0.0435; when Q = 66,500 m^3/s (in 1954), S would be 0.035. The combination of such a high flow and steep gradient is unprecedented among the large rivers of the world. The steep gradient is, in essence, caused by the narrow channel and high composite roughness. Table 1 shows actual roughness data by section.

The section between Fengjie and Wanxian County is wider. When Q is between 110,000 and 66,500 m^3/s, S is between 0.0085 and 0.0095. The Wanxian County–Chongqing section

Table 1

Composite Roughness Coefficient Between Fengjie and Sandouping ($Q = 50{,}000\ m^3/s$)

Section	Roughness coefficient (n)
Taipingxi-Dam	0.0591
Xiangxi-Taipingxi	0.0642
Badong-Xiangxi	0.0733
Wushan-Badong	0.0635
Fengjie-Wushan	0.0661

has three smaller gorges (Huangcao, Mingyue, and Tongluo), and S is between 0.0135 and 0.0145. Actual data show that at Chongqing, when $Q = 110{,}000\ m^3/s$, the water level will be 197 m; and when $Q = 66{,}500\ m^3/s$, 186 m. To protect this major southwestern city, the Central Committee, at its meeting in Chengdu in 1958, agreed to limit the water level at Chongqing after construction of the TGP to 200 m. This is a significant and enlightened decision.

After the completion of the TGP, the composite roughness and flow velocity will be reduced. Past calculations show that the water level at Chongqing would be 201 m when $Q = 83{,}421\ m^3/s$, and 192.83 m when $Q = 63{,}513\ m^3/s$. It is possible that the comrades in charge of the calculations realized that when $Q = 110{,}000\ m^3/s$, Chongqing's water level would be well above the 200 m mark, and thus they did not proceed to do this calculation. This writer believes that in these calculations, the values selected for the composite roughness, initial water level, and sediment inflow have all been too low. Therefore, the water levels that are expected after the completion of the TGP, at 192.83 m and 201 m, may fall far short of the actual values. According to this writer's own estimations based on comparison obtained by using a simplified analysis technique and actual data, Chongqing's water level will be 209.78 m when $Q = 66{,}500\ m^3/s$, and 216 m when $Q = 110{,}000\ m^3/s$. Because this latter figure is 19 m higher than the original calculation, and 16 m more than the limit set by the Chengdu meeting, I consider it to be important enough to request comments from the work units concerned. At the same time, I am conducting more detailed calculations. In any case, comparing the two completed calculations, it is certain that the water level at Chongqing would be much higher than 200 m. The inevitable conclusion is that construction of the TGP would increase artificial flooding in Chongqing. This would be one of the calamities caused by the TGP.

2. The Jingjiang embankments downstream of the proposed Three Gorges dam are located in the most hazardous reaches of the Changjiang River. For infrequent floods in Sichuan province, such as at $Q = 80{,}000\ m^3/s$, the 20 billion m^3 flood prevention capacity proposed for the TGP is adequate to regulate the peak flows, thus reducing the flood hazard along the Jingjiang section. However, the initial trapping of sediment behind the dam may cause scouring at the base of the embankments downstream. This may induce instability and consequently failures at the embankments. A breach may result, and in such an event it will be difficult to organize rescue work. As a matter of fact, slides on the embankments did occur in 1954 at Qijiatuo along the Jingjiang embankments, and at Wangjialiyuan on the Huanghe (Yellow) River. Thus the construction of the TGP has both positive and negative impacts on the Jingjiang embankments. On the one hand, the lowering of the water level at the embankments is favorable for flood prevention. On the other hand, scouring by sediment-free flood flow at the base of the embankments would lead to instability, giving rise to sliding. Hence, with respect to the flood prevention function along the Jingjiang section, the positive and negative impacts of the project are about equal.

3. Hankou is 600 km downstream of the proposed dam site. The 22 billion m^3 flood retention capacity of the reservoir may help to reduce flood hazards along the Jingjiang section, but it is not expected to contribute to flood control at Hankou.

In the middle and lower reaches of the Changjiang, sediment load is progressively reduced downstream (table 2). Sedimentation is dominant in the section between Yichang and Hankou.

Table 2

Sediment Load in Middle and Lower Reaches of the Changjiang

Station	Mean annual sediment load (100 million tonnes)
Yichang	5.2
Hankou	4.35
Datong	4.9

Note: Because of tidal influence, there is no comparable record for stations at the estuary.

However, more than 100 million tonnes of sediment are transported and deposited in Dongting Lake. Thus, some sections of the river bed are scoured, while sedimentation only takes place on flood plains. Recently, professors Ren Meie and Shi Yunliang of Nanjing University presented a graph to confirm this. The graph, reproduced here as figure 1, shows the relation between mean monthly discharge and sediment concentration. The stations upstream (Pingshan, Cuntan, and Yichang) show a clockwise "hysteresis" effect, indicating that sediment is derived from the river basin, whereas the stations downstream (Luoshan, Hankou, and Datong) show a counterclockwise "hysteresis" effect, suggesting that the river bed is one of the sediment sources. It can therefore be concluded that in the initial period of the project, the river bed between Yichang and Hankou will experience erosion and no deposition. After the original equilibrium has been disrupted, enlargement of the channel cross-sectional area will increase the magnitude and frequency of flood peaks and amplify the hazard of flooding in Hankou. The TGP, therefore, creates more flood prevention problems than it solves.

4. The increase of flood hazard in Shanghai is a new problem that may not have been experienced anywhere in the world.

The water level at Wusongkou at the river mouth serves as the datum to measure the severity of flood hazard in Shanghai. On September 1, 1981, the maximum water level at Wusongkou was

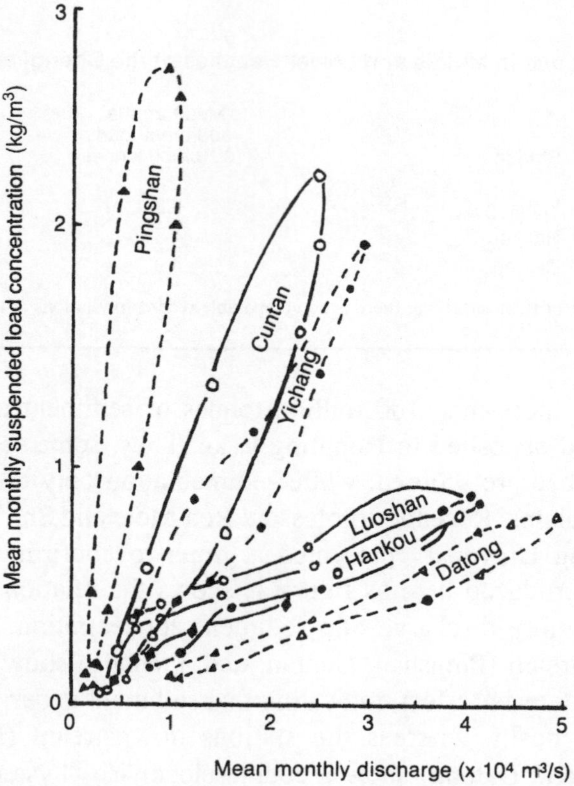

Figure 1. Mean monthly sediment concentration of hydrological stations on the Changjiang.

5.89 m, a record figure. To prevent a further rise of the river and to ensure the safety of Shanghai, tide water was allowed to inundate Pudong. This was unprecedented in the more than one hundred year history of the port of Shanghai.

The Changjiang is a river with a weak tide. There are three sections in relation to the tide: (1) a tide-dominant section extending more than 400 km upstream from the estuary to Wuchao; (2) a salinity-dominant section of more than 100 km up to Xuliujing; and (3) a sediment-dominant section of more than 60 km to

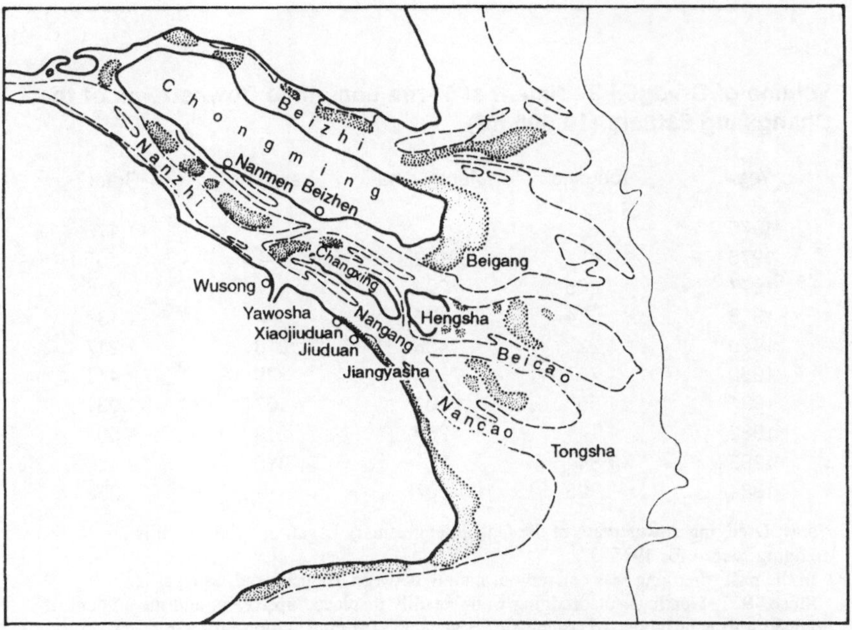

Figure 2. Landforms at the estuary of the Changjiang.

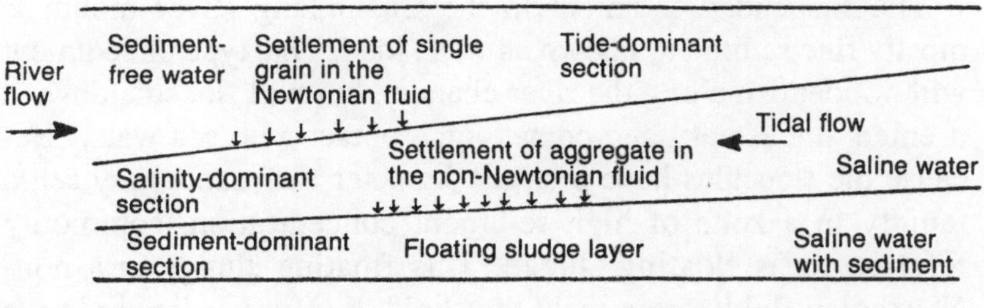

Figure 3. Water and sediment input at the Changjiang estuary

reach Wusongkou (see figure 2). Figure 3 shows the currents at the estuary of the river. The tide- and salinity-dominant sections are well known, but the sediment-dominant section is a peculiar phenomenon observed at the estuary. The conditions at the sediment-dominant section are represented by the amount of sediment dredged from the estuary (table 3).

Table 3

Volume of Dredged Sediment at Three Locations Downstream of the Changjiang Estuary (10,000 m³)

Year	Tongsha	Jiangyasha	Yawosha	Total
1975	1,176		—	1,176
1976	427	142	—	567
1977	605	204	7	816
1978	1,044	139	—	1,133
1979	1,227	436	218	1,217
1980	1,757	337	329	2,417
1981	1,556	207	167	1,931
1982	1,144	260	219	1,563
1983	1,844	271	355	2,476
1984	25	307	—	332

Notes: Dredging downstream of the Changjiang estuary began in 1935; there is no dredging record for 1935–37.

In the past, dredging was carried out mainly between Tongsha and Jiangyasha.

Since 1977, locations for dredging were steadily displaced upstream, and the amount of dredged sediment increased, suggesting a worsening sedimentation problem at the Changjiang estuary.

The suspended sediment in the Changjiang River mouth is mostly fine sediment, known as wash load. This type of sediment will not deposit along the river channel, but will flocculate when it enters the estuary and comes into contact with sea water. Because the floccules have a larger diameter than sand, they settle rapidly in a zone of high sediment concentration, commonly referred to as floating sludge. This floating sludge is a non-Newtonian fluid with a yield threshold. Because sediment in the non-Newtonian fluid may settle in aggregates rather than individual particles, sediment concentration of the hypersaturated fluid continues to increase from the top, reaching as high as 65 kg/m^3. Even at this density, however, it is still a fluid and can migrate with river or sea currents.

Constitutive equations tell us that when an externally applied stress is lower than the yield stress, motion of the fluid is ceased and a pulpy, fluidized mud is formed. Many sedimentologists in

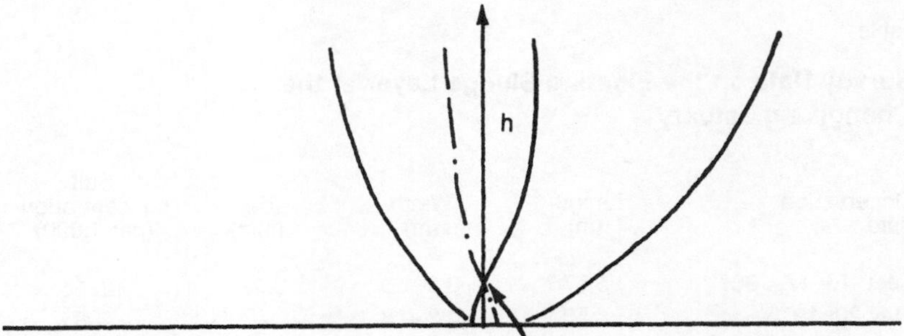

Figure 4. Velocity of flood and ebb tide and stagnation point

the country have done a great deal of work on this process, both in laboratory experimentation and in field monitoring.

Figure 4 shows the distribution of flow velocity at the Changjiang estuary. Surface velocity is higher during ebb tide than during flood tide. During the transition from flood to ebb tide, surface and bottom flow occur in opposite directions. Thus, at the intermediate level, a stagnation point exists where velocity is zero. When stagnation occurs in the floating sludge, the sludge will cease to move, and the resultant formation of the pulpy, fluidized mud may lead to a rapid rise of the water level. A model for the rise of water level at the estuary of the Changjiang may therefore be represented by the following sequence:

Fine sediment → flocculates after coming into contact with salt water → rapid settling of floccules on the saline sediment layer to form non-Newtonian bodies. When stagnation occurs in the salted sediment layer, the floating sludge will cease to move and result in a pulpy, fluidized mud, causing a rise in the water level.

Field observations of the floating sludge at the estuary of the Changjiang have been conducted on a number of occasions. Although there is insufficient information for systematic analysis, the available data are consistent with the suggested mechanism

Table 4

Survey Data on the Floating Sludge Layer at the Changjiang Estuary

Observation date	Length (m)	Width (m)	d_{50} (mm)	Salt concentration (per 1,000)
Sept. 16–17, 1967	5,575	0.6–1.0	0.027	12–18
May 20, 1976	6,500	0.2–0.8	—	5–8
July 6–13, 1976	23,050	0.8–1.2	0.022	6–8
Aug. 3–9, 1976	3,700	0.2–0.6	0.022	6–12
June 10, 1977	5,575	0.2–0.4	0.036	2–9
Aug. 5–10, 1977	8,375	0.4–0.8	0.012	12–22

Source: "Basic Characteristics of River Channel at the Middle and Lower Reaches of Changjiang," October 1983.

Note: According to a paper by Shen Huanting of East China Normal University, the floating sludge layer has been found to be as long as 45 km, with a maximum sediment concentration of 65 kg/m^3.

for formation of the fluidized mud, and they confirm the model described above, especially the requirement that fine sediment and saline water should be present in the floating sludge (see table 4).

The maximum water level at Wusongkou at the estuary is affected not only by peak flood discharge (Q_{max}), typhoon (W), astronomical tides (T), and river bed elevation (E), but also by the floating sludge layer.

In the past, water conservancy workers in Shanghai had thought that the high water level at Wusongkou was due to the simultaneous occurrence of a rising tide (T), typhoon conditions (W), and high discharge (Q_{max}). Results obtained from the analysis of field data suggest that the high water level at Wusongkou is the result of a combination of three factors, namely, a rising tide, typhoon conditions, and a layer of floating sludge (S).

For example, the discharge at Datong in 1936 and 1954 was very large, but the maximum water level only reached 4.48 and 4.49 m, respectively. On the other hand, the peak flood discharge in 1974 and 1981 was not so large, but the sediment concentration was high, and the maximum water level reached 5.29 and 5.89 m, respectively.

Why was it that the extraordinarily large discharges were not correlated with very high water levels? The answer is that these discharges did not have components that generate hyperconcentration of sediment. In fact, the Datong floods in 1936 and 1954 occurred because of high flows both from Sichuan and from the tributaries downstream. The flood water from Sichuan had a high sediment concentration, but downstream tributaries did not. The flood water passing through Datong was therefore diluted. On the other hand, if the flood water at Datong was mainly from Sichuan, and the discharges from downstream tributaries were less significant, then sediment concentration would be high, and water level would be relatively low.

According to observed data, the occurrence of the high water level at Wusongkou above 5 m always coincides with the presence of typhoon conditions. Therefore, typhoon conditions constitute the most important factor for high water level at Wusongkou. The other important factors are sediment concentration (or presence of floating sludge), and a rising tide. The concept that maximum water level was caused by the concomitant occurrence of *W*, *T*, and *S* is developed from the above analysis. According to calculations, the rising tide reached between 4.35 and 4.61 m. That is why it is only the third most important factor.

The last question to be examined is whether the TGP would increase or reduce the sediment concentration at the estuary.

Information from the Sanmen Gorge dam (table 5) reveals that because large particles were trapped behind the dam and finer sediment is discharged, and because of scouring of the river banks downstream, the amount of fine sediment that enters the sea has increased since the completion of the dam. The TGP is more than 1,000 km from the sea. Along this course are collapsed banks. The operating mode of "storing clear water and discharging turbid water" will be adopted at the Three Gorges dam. Figure 5 shows that when Q is higher than 40,000 m^3/s, sediment

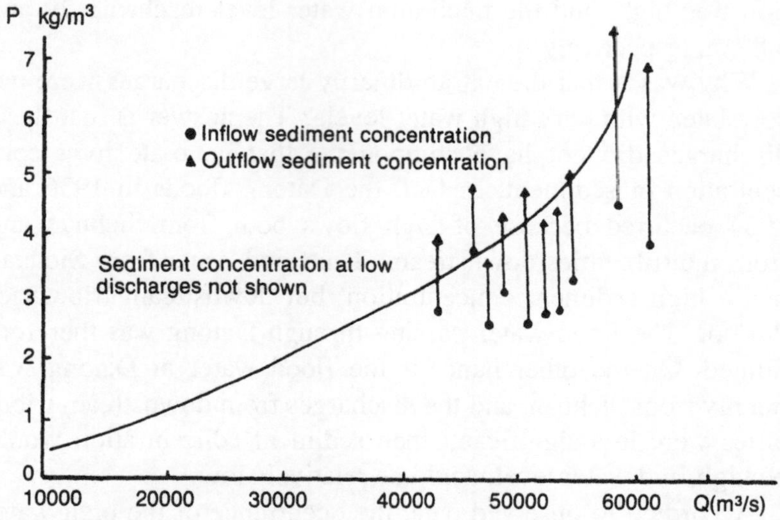

Figure 5. Sediment inflow and outflow for a range of water discharge, Sanmenxia Reservoir

concentration (*P*) of the outflow will rise. All of these suggest that after the completion of the TGP, the amount of fine sediment entering the sea will be increased.

In addition, the graphs prepared by Professor Ren Meie (figure 1) show that for the discharge-sediment concentration relation at Luoshan, Hankou, and Datong a counterclockwise "hysteresis" effect exists. This indicates that sediment in the lower reaches is partly derived from the river bed. After the completion of the TGP, scouring in the main channel will increase the susceptibility of the river banks to collapse. In fact, this process of undermining the river banks is common along the Changjiang. Collapse of the embankments often means that they cannot be rebuilt on the same location but must be built farther away from the channel. *Tuijian*, or "retreat construction," is already a special term connected with the rebuilding of embankments on the Changjiang.

The maximum water level at Wusongkou is a function of *W*, *T*,

Table 5

Data on Scouring at Three Stations Downstream of the Sanmen Gorge Dam during Its Initial Period of Operation

	Discharge from Sanmen Gorge Dam: Effect of retaining coarse and discharging fine sediment at the reservoir			Sanmen Gorge-Xiaolang District: limited scouring and sedimentation — Huayuankou			Lower reaches: scouring of bed and bank — Lijin			
		d < 0.025 mm			d < 0.025 mm			d < 0.025 mm		
Year	Annual sediment load (100 million tonnes)	%	Amount of fine sediment (100 million tonnes)	100 million tonnes	%	Amount of fine sediment (100 million tonnes)	100 million tonnes	%	Amount of fine sediment (100 million tonnes)	Remarks
1961	1.12	90.5	1.014	2.71	33.6	0.91	8.06	48.9	3.91*	*Obtained from survey at Luokou
1962	3.38	21.9	0.74	3.81	54.0	2.06	7.33	58.7	4.53	
1963	6.45	54.0	3.48	5.75	50.1	2.88	9.60	49.2	4.72	
1964	14.80	54.9	8.12	11.90	56.5	6.72	20.3	45.2	8.22	
Average	6.44	51.7	3.33	6.04	51.9	3.14	11.42	50.8	5.60	
Average, 1950–1979							11.00	43.6	4.80	

Source: Data provided mainly by Qi Pu, Yellow River Research Institute.

Notes: It can be concluded that the fine sediment at the estuary is derived from fine sediment discharged from the dam, and from scouring of the river bed and bank.

Compared with the period 1950–1979, it can be seen that the amount and percentage of fine sediment at Lijin have increased since the construction of the dam.

The discharge of fine sediment and the scouring of river bed and bank occurred in a short period of time. This is why fine sediment content at the river mouth has increased. Unfortunately, due to the volume of work involved, no detailed analysis has been conducted.

and *S*. *W* and *T* are beyond the control of man, and *S* will be increased following the completion of the TGP. For the above reasons, average monthly sediment concentration at Datong may reach 1.5 kg/m^3. In that event, according to the empirical relation of *S-H*, the maximum water level at Wusongkou will reach 7 m, and Shanghai will be under a serious hazard of flooding. By contributing to the rise of the maximum water level at Wusongkou, the TGP will be harmful to Shanghai's flood prevention efforts.

Some people believe that the TGP will bring clear water to the estuary, and that this will cause a number of problems. This possibility was demonstrated by the impact of the Aswan Dam, which has caused a substantial reduction in the amount of sediment entering the sea because it has the capacity to store all the sediment. In addition, there are a number of control structures on the Nile, which has no significant tributaries. As a result, this has caused a decline in sardine production at the estuary. However, the Three Gorges Reservoir is different from the Aswan. In the initial period after the construction of the TGP, a situation similar to that on the Nile may arise. To be convincing, however, detailed analysis based on the specific conditions of the Changjiang must be conducted before it can be determined that the same processes will occur.

The TGP will have a significant negative impact on Chongqing, Hankou, and Shanghai, three major cities on the Changjiang. For the Jingjiang embankments, there are both advantages and disadvantages. But overall, there are more disadvantages than advantages. The factors related to the disadvantages are, however, different. At Chongqing, it is the coarse sand and pebble deposited along the channel that will contribute to the raising of the flood stage, retarding the retreat of flood water and prolonging the inundation period. In Hankou, it is the scouring of the river bed along the Jingjiang section, increasing the peak discharge and flood flow at the city. The problem in Shanghai is caused by the increased flocculation of fine sediment, adding to the thickness of the floating sludge layer, and the impact of the pulpy, fluidized mud. These considerations, which are based on information obtained by detailed scientific research, are clearly significant. Hence decisions at the macro level should be made on this basis.

From the perspective of planning for flood prevention, the proposal to proceed with the construction of the TGP cannot be supported. This writer is reminded of his own statement at the 1983 meeting convened to review the project, that the sedimentation issue could become the stumbling block of the project. I only wish that I were mistaken.

From the hydrological perspective, this writer wrote a paper on "Analysis of the Causes of Flooding and the Flood Prevention Measures on the Changjiang." The conclusion was that the TGP should not proceed at the present time. In this paper, I have concentrated on the problems of sedimentation, and again I have concluded that the TGP should not proceed at the present time. I believe that the developmental emphasis on the Changjiang Basin should be on the construction of dams on its tributaries for the purpose of flood retention, sediment storage, and hydropower generation, and the realization of the series of measures already adopted by the Ministry of Water Conservancy and Electric Power. These include: (1) strengthening of embankments, (2) construction of flood diversion schemes in Hunan and Hubei provinces so that each can handle 16 billion m^3 of floodwater when necessary, and (3) investigation of the probability of flood detention where narrow channels exist downstream of Hankou, thus raising the flood prevention capabilities of Hankou. The early completion of these tasks will help to prevent calamities in this vital part of the country, and to ensure that national construction under the four modernizations is not adversely affected.

7

Resettlement Problem of the Three Gorges Project

Fang Zongdai and Wang Shouzhong

WORLDWIDE, a total of forty-nine large-scale reservoirs have been planned for the 1980s. However, as governments begin to realize the adverse effects of dam construction on the natural and human environments, they are changing their policies by scaling down or abandoning altogether their proposals for dam construction. The Soviet Union, for instance, has stopped dam construction planned for the four northward-running rivers to transfer irrigation water to the sprawling plain in the south. Brazil has scrapped its plan to build dams on the Amazon River, while countries such as Malaysia, India, and Australia have all put their dam construction projects on hold.

In December 1985, the U.S. Congress passed a resolution requiring American financial institutions operating worldwide, including representatives of the World Bank, to adhere strictly to environmental protection acts when determining whether or not a project should be funded. Since last year, the World Bank has stipulated that environmental impact assessment must be carried out before dam construction. Mr. Bai Lai Ke Wei Er De [transliteration] of the Institute of Environmental Policies in Washington, D.C., believes that "from an environmental and social viewpoint, the Three Gorges Project (TGP) will cause disastrous consequences."

Fang Zongdai and Wang Shouzhong, "Sanxia gongcheng de yimin wenti." *Daziran tansuo* (Exploration of nature) 4 (1986): 27–30. Translated by Ngan Chu Liu.

All this has prompted us to ask, as we are reassessing the feasibility of the TGP, whether we should analyze soberly the potential disasters and other effects the proposed project would inevitably bring to the natural and human environments. The proposed project will be built along narrow gorges stretching for hundreds of kilometers. If the scheme of a normal pool level of 150 m and a maximum flood control level of 170 m is adopted, hundreds of thousands of people will be displaced by the reservoir, and more than ten medium and small cities will be inundated either partly or completely. These drastic social and environmental changes are unprecedented in the world's history of dam construction. The resettlement problem caused by the TGP is a public issue that cannot be neglected either politically or economically. It must not be taken lightly, and its positive and negative impacts must be analyzed in depth to acquire a complete picture.

1. Four major characteristics of the task of resettlement

1.1. As many cities and towns will be inundated, their relocation and reconstruction are arduous tasks

The 1984 report by the Institute of Urban Planning and Design in China (IUPDC) presents the following findings. The four county towns of Zigui, Badong, Wushan, and Fengjie will be completely inundated. No flat land at higher elevations will be available for relocation of the county town of Yunyang, which is now sited in a section susceptible to landslides. Though less than 50 percent of the county seat of Wanxian will be inundated, the portion affected is the core of the town, which will be disfunctional when inundated. Moreover, the area to which Wanxian can be relocated has been frequently affected by landslides since ancient times. The county seat of Fengdu, about 56 percent of which will be inundated, will be reduced to an isolated island merely 1 m above the new water level. All in all, there are seven cities and counties, namely, Zigui, Badong, Wushan, Fengjie, Yunyang,

Wanxian, and Fengdu, that will have to be relocated and rebuilt altogether. There are three others, namely, Wanxian City, Zhongxian County, and Fuling City, that will be partially inundated and need relocation to higher elevations. The existing built-up area of the above ten cities and counties amounts to 21 km^2 with a total population of 354,000. If the 150-m scheme is adopted, 148,000 people, or 42 percent of the total, will be affected.

Except for Wushan County, all the cities and counties mentioned above received their official designation before the Tang dynasty, more than a millennium ago. Those with a particularly long history include Zigui, which was the former Guizhou; Fengjie, the former Naozhou; Yunyang, the former Yunan; Wanxian County, the former Wanzhou; Zhongxian County, the former Zhongzhou; and Fuling, the former Fuzhou. As these many ancient cities and towns have to be relocated, their cultural relics and historic sites must be preserved or reconstructed. A careful plan should be formulated before construction of the Three Gorges Reservoir. Overall planning and appropriate arrangements made jointly by archaeologists, historians, and departments concerned with such subjects as urban planning and tourism are urgently needed.

Since 1949 a great deal of work has been done on the TGP, but surveys needed to construct topographic maps and obtain the engineering and geological data required for the resiting of cities have yet to be done. Only after IUPDC was assigned to do the work did field surveys begin. The total area required for rebuilding the ten cities and counties was then estimated at 37.7 km^2, or 51 km^2 when taking into account the local topography, which includes unusable gullies and slope land. Once completed, the new built-up areas can accommodate a population of 440,000. However, there are still ten designated towns (including Linjiang, Yunan, Xintian, Wuling, Gaojia, Lidu, Xingshan, and Kaixian) and the Shibaozhai tourist area which are left out of the relocation planning.

1.2. Impoverished land and lack of industrial development in the Three Gorges area pose considerable difficulty for resettlement

For statistical purposes, the Three Gorges area is defined tentatively by the administrative boundaries of the two urban municipalities and seven counties in eastern Sichuan province and the two counties in western Hubei province. The reservoir area is situated between 106°52′E and 111°01′E, and 30°30′N and 31°20′N, bisecting the southern foothills of the Daba Mountains in eastern Sichuan and the hilly area in western Hubei. The area is a narrow strip of land running from east to west. The total cultivated land in the reservoir area is 953.2 million mu, with an agricultural population of 718 million, and a mere 1.33 mu per peasant. Most of this cultivated land is dry slope land with low yield. With an annual income per peasant of merely 232 yuan, this area is a typical poverty-stricken mountain area. Industrially backward, the area is not connected by railways. It is only served by narrow and poor-quality roads winding up and down with the terrain. The Changjiang River is the major waterway, but its water levels vary by 30–40 m between the rainy and dry seasons, and the river conditions change considerably from section to section. Only ships of small tonnage can be accommodated, and port facilities are basically in a primitive state, with all loading and unloading being handled by human labor. The gross value of industrial output of the ten counties and cities was 953 million yuan in 1983, of which Wanxian City and Fuling City accounted for 58 percent, leaving each of the remaining cities and towns with a small share.

There is hardly any agricultural land available for resettlement, and the people relocated will become even poorer. If this problem is to be resolved through industrial development, a large number of medium- and small-scale industrial and other enterprises will have to be established. Otherwise, it will be very difficult to provide the people displaced with employment and livelihood, not to mention improved living standards and increased prosperity.

1.3. Proportion of urban residents displaced is large

The total extent of the ten counties and cities in the reservoir area is about 30,340 km^2, with a total population of 7.79 million. Although the population in the urban centers accounts for merely 4.5 percent, the areas to be inundated happen to be the most densely populated parts of these medium and small cities.

The overwhelming majority of the people displaced by previous large-scale water conservancy projects were peasants, and compensation was restricted to payment for the unharvested crops and relocated houses. The people displaced had to pull themselves up by their bootstraps to earn a living by squeezing into nearby villages and making adjustments to the distribution of farmland. In the case of the TGP, however, more than half of the people to be displaced are urban residents. The costs of compensating the uprooted settlers involve, first, the cost of housing construction. This leads to the problem of funding for building houses and the necessary public facilities, such as water and electricity supplies, roads, and related urban infrastructures, facilities for culture and sports, education, and hygiene, service industries, land and water transport with the outside, and communications. In the case of Tangshan Municipality, the government's special allocation and locally raised funds for redevelopment after the earthquake totaled 8.041 billion yuan for the past ten years. On the basis of the 73 km^2 of Tangshan's urban district and the 226,000 households, the unit investment per km^2 in urban facilities was as high as 110 million yuan. It should also be noted that with its originally concentrated built-up area and level terrain, the city basically underwent redevelopment at its original site. This is clearly different from the case of the TGP, in which a dozen cities and counties will have to be relocated and rebuilt from scratch. Considering the rugged topography of mountain land and deep gorges, winding roads, and effects of the fluctuations of water levels on revetments and piers, the volume of construction and expenditure involved in resettlement will have to be larger.

The costs of resettling urban residents are too large to be compared with those necessary to relocate the peasants. Estimates of resettlement costs in the Xiaolangdi Reservoir area on the Huanghe (Yellow) River suggest that the costs of resettling each person engaged in a rural and township enterprise is about 6,000–7,000 yuan; in light industry, 10,000–12,000 yuan; and in heavy industry, more than 20,000 yuan. The cost of resettling an urban resident is thus three to four times higher than resettling a peasant.

1.4. Area inundated and number of people displaced are closely related to water level in the reservoir

The Three Gorges are deep canyons lined with steep mountains on both sides, and the densely populated areas lie between 125 m and 210 m above sea level. According to the statistics of the Planning Office of the Changjiang River Basin (POCRB), a normal water storage level of 150 m will displace a total of 330,000 people; 160 m, 420,000 people; 170 m, 620,000 people; and 180 m, 880,000 people. But these figures are far from the truth.

In the case of the 150-m scheme, the number of people to be displaced estimated by the affected cities and counties is larger than that by the Urban and Rural Planning and Design Institutes in Hubei and Sichuan, whose estimate again exceeds that of the IUPDC. The figure by the IUPDC is, in turn, larger than that by the POCRB. All of these estimates have not yet taken into account the costs of establishing enterprises required for resettling the urban residents.

The data provided by the IUPDC are presumably more reliable as they are obtained through field surveys and statistical analysis. The summary report of the outline plan prepared by the IUPDC, however, has not yet been seriously examined and approved by the departments concerned. Moreover, once the Three Gorges Dam is completed, siltation at the tail of the reservoir will increase and the level of backwater profile will rise each year. This will undoubtedly enlarge the area to be affected and increase the size of population displaced. Another problem that has to be resolved concerns the fact that the TGP is indeed a high dam for low [water storage level] use. A semiflood zone with an altitude of 20 m will be created when the reservoir is impounded between the normal pool level of 150 m and the maximum flood control level of 170 m. Whenever the excess water is stored, residents in the semiflood zone will have to flee, and houses and farmland will be destroyed. Controversy still surrounds this issue. Nonetheless, it is certain that flood damage will occur, and compensation is necessary.

The special characteristics of resettlement caused by the TGP make it a more formidable task, and one that should not be taken lightly by concerned departments. While a resource-exploiting type of resettlement policy is undoubtedly a correct answer to the problem, many specific issues must be dealt with in a realistic and serious fashion to avoid problems in the future.

2. The task of resettlement is crucial to the success of the TGP

The construction of the Guanting and Sanmenxia reservoirs during the early period of the People's Republic have left many resettlement problems unresolved even until today, due to the former "leftist" policy of overemphasizing the interests of the whole at the expense of those of the individuals concerned. Although the reservoirs are completed, the reservoir areas remain as poor and backward as before. It is true, however, that the problem of resettlement cannot be solved by going to the other extreme, setting premature and unrealistic standards regardless of China's financial capability. That would only gain a little by losing a lot, hampering the implementation of water conservancy projects, which are of importance to the entire country. Therefore, the first and foremost task is to formulate resettlement policies suited to the Chinese conditions. These policies should take into account not only China's financial capability and conditions, but also the need for improving the living standards of the displaced population so that they can see their stake in the process.

After all, resource-exploiting resettlement is an experiment in China. To carry this out, as some suggest, two issues need to be resolved: first, instead of being required to be self-sufficient, the people to be displaced should be provided with commodity grain by the state; second, passive resettlement should be transformed into planned development. Instead of being provided with employment opportunities in agricultural production alone, the displaced population should be given a wider choice by changing the production structure of the reservoir area. In other words, a batch of small- and medium-sized industrial and other enterprises should be established in the reservoir area according to local resources for the people to be displaced.

While it might be relatively easy to provide the hundreds of thousands of resettlers with commodity grain, it might not be so easy to resolve the second issue. The establishment of small- and medium-sized industrial and other enterprises requires not only capital, raw materials, technology, and auxiliary facilities such as water and electricity supplies and transportation, but also constant improvement in the quality of the labor force—in education, knowledge, skills, and ability to ensure that the enterprises are competitive and able to produce high-class and marketable goods. The sales and marketing of commodity goods are fundamental to raising the living standards of the people resettled after they have changed their types of employment. This kind of "planned development" policy should proceed scientifically from the start: we should carry out surveys on the resources in the reservoir area; formulate specific development plans for the enterprises; and carry out pilot studies after sufficient socioeconomic surveys are completed. Only after experimental studies are successful should the above policy be widely implemented.

Just what ecological and social effects the TGP would cause is another issue that should be taken seriously. Most of the small and medium cities and towns in the reservoir area, for instance, are located, for historical and economic reasons, on the north bank of Changjiang River. According to the siting surveys by IUPDC, some cities and towns will be relocated south of the river due to topographical constraints. However, as the south bank is backed by steep mountains, will it receive less sunlight than the north, and will it be harmful to the human living environment? Also, because it is made up of sandstone, limestone, and shale, the Three Gorges area is susceptible to landslides after prolonged rainfall, and this has already happened many times in the past. After completion of the reservoir, the frequent rise and fall of the water level will invariably exacerbate the perils of landslides, causing loss to the residents. A solution therefore must be found with respect to the area's geological structure; otherwise, this potential threat will only create concerns and skepticism about the tasks of resettlement and urban reconstruction.

As the proposition of resource-exploiting resettlement is still at a relatively rudimentary stage, it will certainly require a long and arduous process to move from the various pilot studies to full implementation of the concept. From the perspectives of sedimentation and urban and rural construction, and based on fragmented information gathered so far, we have written this short article, hoping that all concerned will investigate these issues in more depth.

Note

1. *Jianzhu dongtai* (Developments in construction) 30 (July 1986).

8

Population Resettlement and Economic Development in the Three Gorges Reservoir Area

Tian Fang and Lin Fatang

THE RESETTLEMENT of the population displaced by the Three Gorges reservoir and the economic development of the reservoir area are important issues of both principle and practice in the Three Gorges Project (TGP). They have aroused concerns in various quarters. Surveys are being actively conducted and resettlement plans proposed. Solving these major problems requires the following: an appropriate assessment of the economic impact of inundation on the reservoir area; a thorough understanding of the advantages of natural resources and development prospects enjoyed by the area; and a proper management of the relationship between resettlement of the people displaced and the development of the regional economy. If these problems are resolved, the TGP will serve as an advanced water conservancy project, and it will bring prosperity to the reservoir area. It will undoubtedly become a model for emulation in the course of China's four modernizations.

Tian Fang and Lin Fatang, "Sanxia shiuku de yimin anzhi yu kuqu jingji fazhan." *Zhongguo renkou qianyi* (Chinese population resettlement), pp. 199–210. Translated by Pingnan Yuan.

Benefits of the project and losses caused by inundation in the reservoir area

For any water conservancy project, the benefits are gained at the expense of losses caused by inundation in the reservoir area. There is no doubt that the tremendous benefits of the TGP in flood prevention, hydropower generation, transportation, and so forth are obtained at the expense of substantial sacrifices by the people in the area. For a normal pool level of 150 m (which is under review), the areas that will be constantly flooded and affected by backwater will include all the towns and villages in fourteen counties and ten cities along the banks of the Changjiang River in Sichuan and Hubei. The population to be resettled will be more than 300,000. (If the population increase from the countryside and cities and the return migrants are taken into account, the actual number of people to be resettled would be 460,000 by 1994). Farmland inundated will amount to more than 100,000 mu. However, the inundation index, that is, population to be resettled and farmland inundated per kW of hydropower generated, is far lower than the average of other large- and medium-sized hydropower stations already completed or under construction in China. The number of people to be displaced is roughly equal to that of the Xinanjiang (290,000), Danjiangkou (380,000), and Sanmenxia (320,000) projects.

For the TGP, not only is the inundation index favorable, but the distribution of displaced people is scattered. The proportion of population requiring resettlement in the various counties and cities is small, and the absolute number of peasants to be resettled is omitted. As the reservoir only occupies the main river channel and extends into the tributaries, people displaced and land inundated will be distributed along the banks of the 1,000-li reservoir area from Sandouping to upstream of Fuling. Taking as an example Wanxian prefecture, an area with the largest number of people to be resettled, only 38 percent of the population to be displaced, or 80,000 people, are from agricultural households. They are scattered in 2,168 production teams, which belong to

503 brigades or 104 communes, and 41 designated towns and urban districts within the jurisdiction of seven counties and cities. Besides, there will only be a few totally inundated brigades and production teams. This will allow the displaced population to "leave the land but not the village," and still resettle within the same general area but at higher elevations, thus involving relatively less conflict in the use of land.

The management of losses resulting from inundation will not only have to take account of the private deprivation and other immediate interests of the population needed to be resettled, it will also need to consider the effect on industrial and agricultural development and the direct loss of revenue of the affected region. This is because, as a region along the Changjiang River and its tributaries, the reservoir area enjoys convenient transportation, supports a concentration of urban centers with a dense population, a developing industrial economy, and prosperous regional commerce. According to surveys in Yunyang County, Sichuan, an area to be heavily flooded, the land inundated will only account for 2.5 percent of the county's total area; the people displaced, 3.7 percent of the county's total population. However, the inundated area bears 67 percent of the county's total population in designated towns and rural markets, 88 percent of its fixed industrial asset, and 67 percent of its existing installed thermal and hydropower generating capacity of 24,000 kW. Therefore, the regional economic impact of inundation caused by the reservoir cannot be neglected.

The TGP, which will generate tremendous benefits and which is favorably located and of strategic importance, cannot be replaced by other projects. Besides, maximum economic benefits will be obtained with minimum losses caused by inundation. However, it should be realized that the scale of the losses caused by inundation is large, involving many departments and exerting a wide impact. In addition to the resettlement, there will be an indirect economic effect on the adjacent areas of the reservoir, where the original socioeconomic conditions may be changed. Moreover, there are still many problems that are difficult to express in monetary terms. Hence, a correct policy of population resettlement should be adopted, a pragmatic master plan for resettlement formulated, and active measures taken to protect the interests of both the people displaced and those already in the resettlement areas. In this way, as the TGP proceeds, the economy of the reservoir area will gradually develop, and the people in the area will become increasingly prosperous.

Population resettlement and development of the reservoir area

The TGP is composed of two closely related components, the project structures and the reservoir. The coordination of resettlement with construction of the structures and the development of the regional economy by taking advantage of its resource base are not only strategic tasks of profound influence, but also bear important practical significance.

Developing the reservoir area for population resettlement

Inundation caused by the Three Gorges Reservoir will lead to resettlement of hundreds of thousands of people, relocation of tens of cities and rural towns and hundreds of factories, and reconstruction of a variety of special facilities. Productive and nonproductive sectors that need to be established in the vicinity of the resettlement areas include the following categories: industry, agriculture and forestry, water conservancy, transportation, urban and rural construction, environmental protection, commerce, culture and education, hygiene and medical care. Such a complicated and difficult task must be conducted on the basis of developing the reservoir area and its economy, and from the perspective of systems engineering, which will contribute to the selection of a cost-effective and practicable scheme.

The problem with resettlement of peasants, who depend on farming as a livelihood, essentially concerns the employment of the labor force. However, there are already too many people utilizing a limited amount of farmland in eastern Sichuan and western Hubei, where farmland area per capita is only about 1 mu. As a result of the tremendous increase in productive efficiency due to agricultural modernization and the adoption of the contract responsibility system, 30 percent of the present labor force has become surplus and requires new jobs. Consequently, it is no longer possible to find new farmland in the area to absorb large numbers of migrants. Hence, a strategic shift to exploit the surrounding mountainous areas is necessary. This should be conducted in conjunction with an adjustment of the structure of agricultural production, full exploitation of resources, and initiation of new production opportunities.

Based on the actual situation in the Three Gorges Reservoir area and preliminary surveys and planning, a production structure of diversified economy and comprehensive development may be established in the course of building up the resettlement area. Within this structure, several approaches may be adopted: reasonable division of labor; simultaneous advancement of the various production sectors; joint management of agriculture, industry, and commerce; promotion of commodity production; processing agricultural and subsidiary products in situ; refinement and comprehensive utilization of products; and promotion of the utility ratio and economic value of products. The culmination of these approaches may see the reservoir area providing more products to the country and gaining higher income, thus enabling both the resettled people and the original residents to develop fully their potential and to obtain mutual benefits.

The relocation of cities, towns, and factories must be supported by the vast areas in the vicinity of the reservoir. Cities and towns have their areas of influence and serve as political, economic, cultural, transportation, and commodity distribution centers. If they are relocated outside the reservoir area, they would lose their raw material bases, markets, and conditions for existence. There is therefore a close relationship between the reservoir area's economic development and the nature, size, development direction, population structure, functional zoning, industrial distribution, public construction, and urban facilities of the cities. Moreover, economic ties between the cities should not be broken due to relocation, for this will lead to reduced commercial benefits. Rather, economic development in rural and urban areas should be coordinated. There should be a rational division of specialized functions among the cities, which in turn should form an integrated, organic, and modern urban system as well as a harmonious economic unit in the Three Gorges area.

Abundant resources to be exploited

The Three Gorges area is located in the heartland of China. Upstream of the reservoir area is the well-endowed Sichuan province, and the land of fish and rice in Hunan and Hubei provinces are located downstream. The Three Gorges area possesses exceptionally rich natural resources and is strategically located. Permanent and temporary inundation affects nineteen counties and cities which have a total population of thirteen million and an administrative territory extending more than 50,000 km^2. The area affected may be even larger if the traditional economic connections and current commercial ties are considered. With its charming scenery, abundant natural resources, moderate climate, and plentiful precipitation, the Three Gorges Reservoir area and its vicinity provide favorable conditions not only for a comprehensive development of agriculture, forestry, animal husbandry, fishery, and sideline production, but also for raw material extraction and processing, transportation, and tourism.

Population resettlement and the economy of the reservoir area are interrelated

The resettlement problem due to inundation in the Three Gorges Reservoir area cannot be considered as an isolated issue but should be related to the socioeconomic situation in the resettlement area. The vigorous economic development of the reservoir area provides the possibility for resettlement and its further consolidation, while the resettlement brings about a favorable environment for the establishment of the "two civilizations." Proper coordination and planning must be done to allow resettlement and economic development of the area to take place while the TGP is being constructed, to ensure that the living standards of both the people resettled and those already in the resettlement areas are truly raised.

The tasks of resettlement and of developing the economy of the reservoir area should be and can be integrated. On the one hand, construction to prepare for inundation and resettlement will need resources that will be purchased by the state. On the other hand, the abundant natural resources in the reservoir area have hitherto not been fully exploited because of the lack of investment and technology. It is possible to integrate the two through a unified plan implemented in stages during the course of the construction of the TGP. The link between resettlement and development of the economy of the reservoir area is the collective sharing of benefits. According to China's current economic policies, the compensation for damages due to inundation is tentatively estimated at 2–3 billion yuan. This is seven or eight times the combined revenue of Wanxian, Fuling, and Yichang prefectures in 1983. Annual expenditure is estimated at two to three times the total investment in capital construction by state-owned enterprises in these three areas. The investment in urban and rural construction for resettlement, in particular, will be mainly used for the purchase of local materials and for wages, thereby facilitating individuals and local governments to increase their income and capital accumulation. The economic benefits can be extended into broader areas to raise the income for both the people to be resettled and those already residing there, and to lay the foundation for long-term development. This will become a reality provided that we follow objective laws, improve production structure, balance production relations, and deploy a large volume of investment in the development of the reservoir area and in commodity production.

The key to economic success in the reservoir area

The problems of inundation management and organization of resettlement are interrelated in many ways. Thus, in addition to considering the resettlement and development of the reservoir area as an organic entity and conducting systematic studies and careful planning to achieve concrete results, the critical issues for the success of the Three Gorges Project are to strengthen unified leadership, to formulate correct policies and practicable measures, to utilize favorable conditions, and to mobilize various positive factors.

Restructure the management system and simplify the administrative hierarchy

The areas to be inundated, including areas to be temporarily flooded, cover parts of five prefecture-level cities in Sichuan and Hubei. People from all walks of life will be affected. In the case of the relocation of industrial and mining enterprises, administrative authorities ranging from the central government to the provincial, prefectural, county, district, village, and production team levels are involved. There are simply too many authorities with different interests and requirements. If the current management system and administrative hierarchy remain unchanged, maintenance of the unity of the Three Gorges regional economy will not be provided. Historical mistakes such as uncoordinated management, duplicate development, wasteful use of volunteer labor, and limited funds may be repeated.

To mobilize the residents in the reservoir area to exploit the natural resources, develop the economy, promote resettlement, ensure that the TGP can be successfully completed, and bring all the benefits into full play, a unified and more simplified chain of command must be established according to the present situation and positive experiences in China. Under such a command, excessive administrative units should be eliminated. The county governments should be streamlined and staffed with capable personnel. These counties should be administered by the city governments of Yichang, Wanxian, and Fuling, within a structure that eliminates the prefectural level. Cadres who have become surplus as a result of this simplification should be retrained and transferred to manage the various enterprises.

Adopt special policies and speed up construction of the reservoir area

Success in resettlement and economic development in the reservoir area depends on the adoption of proper policies and a scientific approach. Following the principle of "opening to the outside world and revitalizing China's economy" and proceeding from present conditions, the following policies should be adopted:

1. Expand the jurisdiction of the reservoir area. Following the principle of centralized leadership and planning, the reservoir area should be allowed to develop projects according to its local conditions, to export local goods directly overseas, to manage by itself the tourist industry, and to give foreign investment preferential treatment.

2. Tax reduction and remission, and preferential treatment. Low-interest loans and extended repayment periods should be granted to such industries in the reservoir area as agriculture, mineral extraction, fisheries, animal husbandry, food processing, transportation, construction and building materials, electronics, and tourism. Also, a basic fund for development should be established to promote public services such as culture and education, medical care and hygiene, broadcasting and television, and postal services. Taxation should be reduced to one-half during the period of construction. Incentives should be provided for the development of new technologies such as computers, communications, materials, and biological engineering.

3. Raise capital from various sources. At present, the central government provides a subsidy of 100 million yuan per year to the prefectures in the reservoir area. Compensation provided for inundation is only sufficient to cover the cost of resettlement and of rebuilding some facilities. It will not be able to satisfy the

capital requirements for regional development. Thus, capital should be raised from as many sources as possible. Part of the compensation funds should be provided in advance. Annual subsidies and the regular budget should be paid three to five years in advance into a development fund. Foreign and domestic investment should be attracted through joint ventures and compensatory trade. Shares can also be sold to enterprises, collectives, and individuals. In addition, a certain proportion of the benefits derived from hydropower generation should be plowed back for resettlement purposes. Industries that require low investment and have a short capital cycle and a high return should receive the highest priority for development in order to accumulate capital.

4. Grain purchase policy. Exemption or reduction of food grains procured by the state should be the policy in the resettlement areas. The government should provide food grain to the resettled people through rationing. The policy of equalizing benefits should be adopted to ensure that the grains required by areas upstream of the reservoir are supplied at regular prices by the downstream areas that reap the benefit of flood prevention due to the construction of the reservoir. The reservoir area should be allowed to trade its oranges, local specialties, and other goods for food grains from grain-producing areas.

Provide adequate preparation for resettlement and emphasize economic efficiency

The development of the reservoir area should be premised upon bringing its advantages fully into play and avoiding its shortcomings. The policy of opening to the outside world and activating China's economy, the construction of the two civilizations, and the project itself should be coordinated to achieve the best economic result and to help the people in the reservoir area to become prosperous quickly. Centralized leadership is needed to direct the intricate tasks of resettlement because of the large size of the affected area and its varied natural and social conditions. Work must proceed according to the principles of systems engineering and through careful planning and rational coordination, to obtain the anticipated results with minimum costs.

The issues of resettlement and economic development of the reservoir area are not only related to the rational utilization of local resources but directly affect the revitalization of the regional economy and the immediate interests of the many households and individuals. This is a situation involving interdisciplinary studies in the natural and the social sciences and requiring strong policy support. It is much more complicated than the situation normally associated with development in an economic area or a key project. Only through careful investigation can a pragmatic blueprint of construction with rational alternatives and scientific prediction be developed. Advanced technology and management experience should be actively diffused to ensure that all the undertakings proceed from a scientific base, so that optimum socioeconomic results are obtained with the minimum investment of money and time.

To ensure the return of investment, it is necessary to establish a system that lays out a clear division of responsibility, rights, and benefits and strictly requires adherence to the procedures of capital construction. Once approved, construction units, whether they be industrial enterprises, urban and rural structures, public utilities, or agricultural infrastructure, should be built according to the design specifications. The state bank should monitor the expenditure on each unit. The authorities in charge of the construction should sign contracts with the groups undertaking the construction tasks, and the local urban or rural organizations. These contracts should be certified by judicial departments as the legal bases for implementation and scrutiny. Cadres and workers cannot be personally appointed by an individual and receive life-long employment. They should be employed according to their qualifications. While investment in education should be emphasized, those cadres who mismanage factories and those enterprises that have recurrent deficits should be dealt with according to regulation. If all of the above are followed, it will be possible for the reservoir area to develop rapidly and efficiently, from a condition of underdevelopment to become one of the most prosperous, advanced, and attractive developed regions of China.

9

How Much Investment Is Required by the Three Gorges Project?

Wang Jiazhu

THE CONSTRUCTION of the Three Gorges Project (TGP) needs exactly how much investment? What are the economic benefits of such investment? How can such investment be recovered? These are vital issues that must be resolved. They are also some of the focal points of controversy between the proponents and opponents of the TGP.

Cost effectiveness indicators of the TGP

The preliminary design report formulated in March 1985 by the Planning Office of the Changjiang River Basin (POCRB) of the Ministry of Water Conservancy and Electric Power calls for the construction of a high dam of 175 m for a normal pool level of 150 m, and a hydroelectricity generation station having an installed capacity of 13 million kw. An estimation using the country's 1984 price levels and following the rules and regulations in effect shows that the total investment would be 15.947 billion yuan, made up of 12.4 billion yuan for the key installations and 3.547 billion yuan for compensation and resettlement

Wang Jiazhu, "Sanxia gongcheng jiujing xuyao duoshao touzi." *Diqu fazhan zhanlue yanjiu* (Research on regional development strategies) 3 (1987): 40, 43–46. Translated by Jane Tan.

of people affected by the creation of the reservoir (see table 1).

Because the TGP provides enormous benefits for flood prevention, hydroelectricity generation, and navigation, although the total investment is large, the benefits per unit of investment are high. After taking note of the formula of cost sharing among the different parts of the project and the state's stipulation that it does not expect any financial return from its investment in the flood prevention component of the project, either in the form of repayment of the principal or as interest on the principal, the per kW investment is only 858 yuan, and the per kWh of electrical energy investment is 0.165 yuan. These are not only lower than the averages of the other large-scale hydroelectric generating stations currently under construction (their per kW investment being 1,500 yuan, and per kWh of electrical energy investment 0.35 yuan), but also close to those of the thermal generating stations. In addition, hydroelectric generating stations enjoy many significant advantages over thermal plants. The former have low operating costs, do not require additional investment in building coal mines and transportation facilities, and are nonpolluting. Even if the cost-sharing formula is omitted from consideration, that is, if the full amount of investment were to be recovered from the hydroelectric generating sector, the per kW investment would only be 1,230 yuan, and the per kWh of electrical energy investment 0.236 yuan, still lower than the averages of the large-scale hydroelectric generating stations currently under construction or proposed.

Some comrades who hold a different view of the TGP have different interpretations of the above statistics. They point out that calculation of total investment should not be limited only to the costs of the project, but should also include expenses incurred in the provision of related installations such as transmission lines. They also suggest that price increase and loan interest be taken into account. We agree that to be compatible with the new situation brought about by recent economic reforms, it is both reasonable and necessary to adopt a dynamic rather than static approach to economic analysis to assess the rationality of the project. But

Table 1

150-m Normal Pool Level: Estimates of Investment (100 million yuan)

Permanent installations									
Dam	Hydro station	Navigational facilities	Other	Subtotal	Temporary installations	Other installation costs	Preparation cost	Compensation and resettlement	Total
13.73	39.78	13.32	6.59	73.42	23.34	19.13	8.11	35.47	159.47

we do not think it is a scientific comparison and evaluation when some use the total investment of the TGP derived according to the above dynamic method to compare with investments in other projects calculated in the usual static fashion.

Number of years needed to recover total investment

In view of the TGP's large size and long completion period, during the early design phase a method that took into consideration the time factor by discounting the costs and benefits was adopted. Comparison was made between the Three Gorges hydroelectricity generating station and alternative hydro and thermal plants with similar performance. The results show that at a 7 percent annual interest rate, when the electricity price is above 0.04 yuan per kWh, the internal rate of return will be higher than 14.2 percent (when the cost-sharing formula is considered), and 10.6 percent (when the formula is left out). This is more beneficial than the alternative scheme, and superior to the current average investment efficiency of the whole electrical power industry. Using the same principles for calculation of the recovery period, it is found that it takes 2.7 years after construction to recover the total investment (when the cost-sharing formula is considered), and 7.9 years (if the formula is left out). [Note that these values are far more optimistic than those indicated in the introduction.—eds.]

The superior cost-effectiveness indicators of the TGP are due to the exceptional natural advantages enjoyed by the Three Gorges: abundant water, beneficial landforms, and favorable geological foundation. The more than thirty years of survey, design, and research work after the establishment of the People's Republic have given the project ample and reliable information about the hydrology and provided a clear picture of the favorable geological conditions. The locations of hydrological and engineering structures are determined on the basis of a long process of selection. All of this preparatory work has provided assurance for maximum project benefits and placed the calculation of engineering work on a comparatively reliable basis. It is unlikely that major changes in the calculation will be needed, and the basic precondition for reliable investment estimates is thus ensured. Table 2 shows a comparison between the physical indicators of the TGP (150-m normal pool level) and those of the hydroelectricity plants under construction or proposed. It can be clearly seen that the amount of concrete engineering work per kW on the TGP is much lower. Even though many opponents of the TGP have persistently emphasized the great losses due to inundation, the project's per kW inundated area and population requiring resettlement are again lower than the comparative figures for other plants.

The large size of the project and its long completion period no doubt create disadvantages for the recovery of investment. After long years of study, in the preliminary design it was recommended that this situation be improved by building a three-phase cofferdam to enable an early start for hydroelectricity generation. The overall schedule calls for three years of preparatory work, three years for the first phase of the project, five years for the second, and six years for the third, the project completion period being seventeen years. Using the above short-cut, it is possible to make the first two groups of turbines generate 1 million kw of electricity in the eleventh year, when investment has reached 64 percent of the total. Thereafter, each year four groups of turbines producing 2 million kw can go into production. This arrangement, allowing for production of benefits while construction is still going on, can create more than 270 billion kWh of electricity and a gross output value of 18 billion yuan (at a per kWh value of 0.065 yuan) over the whole construction period. No other generating plant can match such superior performance. Comrades in the central leadership have often pointed out that the key to investing in large-scale hydroelectric plants lies in the amount of investment made prior to the commencement of production of the first group of turbines. One characteristic of the TGP is the larger proportion of investment in electrical installations and resettlement. This helps to reduce the investment necessary before

Table 2

Comparison of Physical Targets for Three Gorges Hydroelectric Plant and Other Hydroelectric Plants

Item		Installed capacity (10,000 kW)	Inundated cultivated land (10,000 mu)	Population to be resettled (10,000)	Concrete work (10,000 m³)	Inundated cultivated land (mu) per 10,000 kW	Population resettled per 10,000 kW	Concrete work (10,000 m³) per 10,000 kW
Three Gorges hydroelectric plant		1,300.00	14.61	33.04	2,291	112	254	1.76
Hydroelectric plants completed or under construction	11 completed plants in East and South China	381.20	139.90	120.99	1,219.73	3,670.00	3,173.92	3.20
	7 plants under construction in all China	669.50	28.09	21.76	1,973.00	419.57	325.00	2.95
	Subtotal	1,050.70	167.99	142.75	3,192.73			
Proposed hydroelectric plants	70 in Central and East China	1,488.13	248.42	204.00	4,711.20	1,669.34	1,370.85	3.17
	6 in Sichuan and eastern Guizhou	769.00	18.75	24.37	2,362.30	243.80	316.90	3.07
	Of which 4 (Wuqiangxi, Geheyan, Shuikou, Pengshui)	510.00	13.88	23.82	1,231.00	272.20	467.10	2.41

the first group of turbines is put into production. Thereafter, because there is a large output of electrical energy, the investment problems are not difficult to resolve.

Validity of Gezhouba Project investment estimate suggests credibility of calculations derived through regular design process

In discussing the investment and benefits of the TGP, it is useful to review the investment and benefits of the Gezhouba (Gezhou Dam) Project and summarize the experience and lessons.

The Gezhouba Project involved the construction of the first large dam on the Changjiang River. At the completion of the first phase of the project, its technical achievements received the attention and praise of the hydroelectrical engineering community both at home and abroad. But there has not been much discussion about the economic benefits of the project.

According to the plans for the Changjiang Basin, Gezhouba is supposed to act as the navigational step downstream of the TGP, to improve navigation in the 40-km stretch of canyon below the high dam, and to serve as a counterregulation reservoir to solve the problem of irregular daily discharge at the Three Gorges Dam. The primary purpose of Gezhouba, therefore, is to improve navigation. This means that its benefits can only be fully realized after the completion of the TGP.

Gezhouba's construction experienced a winding path. In name, the first phase began on December 30, 1970, at the height of the Cultural Revolution when the two antiparty cliques of Lin Biao and Jiang Qing were on a rampage, and when ultraleftist ideas were spreading unchecked. The views of the technical personnel were suppressed. Engineering work was hurriedly commenced without ample survey, design, and research, in the absence of a single responsible work unit, and before a series of major technical problems were resolved. By 1972 it had become impossible to continue work on the project. Standing against the strong pressure from the Gang of Four, Premier Zhou Enlai determined that the construction of the principal part of the project be halted after he was fully briefed on the progress of the project. He also ordered that the Gezhouba Project Technical Committee be established and appointed the POCRB responsible for making the necessary changes in the project. In the two years following this, cooperation was established with the country's universities and colleges and research, manufacturing, and construction units to resolve satisfactorily the series of major technical problems. Work was resumed in October 1974. After the Gang of Four was crushed, progress accelerated. The first phase was basically completed by mid-1981, and the project began to produce benefits for navigation and hydroelectricity generation. By 1986, the second phase hydroelectricity generation also began. The project was originally scheduled to be completed by 1989, but it seems from its current progress that this can be brought forward by one year.

The State Planning Commission has formally examined and approved two sets of estimates of costs. The first set, approved in July 1975, showed a total budget of 3.556 billion yuan (of which 2.3 billion yuan was for the first phase); the second, approved in May 1984, indicated an amended total budget of 4.848 billion yuan. Major reasons for the increase included the following: the scale and performance of the second phase were raised (installed capacity increased from 2.215 million kW to 2.715 million kW; transformer output voltage was raised from 330 kV to 500 kV; the sediment sluice gate was also enlarged; etc.); changes in state policy; adjustment of rates; and price escalation. In passing, it may be pointed out that it is unfair for some to take the 1.35 billion yuan cost estimate made at the preliminary stage of construction work as the total budget and blame the design unit for this. As mentioned earlier, the project was commenced at an abnormal time, during the Cultural Revolution, and was done without the necessary survey, design, and research work to serve as basis. A series of major technical problems had not been resolved, and no work unit assumed overall responsibility for the project. Under such circumstances, design personnel had to follow the dictates of whomever was in power. Estimates derived this way cannot be compared with those that have gone through the formal process of design and documentation.

The actual spending of the budget for the Gezhouba Project is as follows. For the first phase, the 1975 estimates amounted to 2.3 billion yuan; the approved 1984 statement of accounts showed a total expenditure of 2.471 billion yuan, representing an over-expenditure of 7 percent. A breakdown of the total expenditure shows that under the permanent structures item there was a surplus of 200 million yuan, but a deficit of 150 million yuan occurred under the temporary structures item. Another increase of 280 million yuan was due to the change in state policy and price increase. After analysis, it is concluded that the 1975 estimates were basically reasonable. If stricter control were exercised in the course of construction, expenditure could have been kept within the limits of the budget. In the case of the second-phase estimates, up to the end of 1986, 1.84 billion yuan had been spent, leaving an unspent sum of 530 million. Preliminary calculations show that the total budget of 4.848 billion may not be exceeded. During the last few years, due to the reforms the country has gone through, price and policy changes have caused significant increases in project expenditures. However, it is possible to keep expenditures within the estimates due to the reliable information about the project on which the estimates were based, and the larger contingency fund provided under the estimates. The contract system adopted during the second phase also improved the management of construction work and gave better control over spending. This indicates that estimated figures derived through the regular design process are credible and acceptable.

Since the completion of the first phase in 1981, benefits have been derived from the Gezhouba project while construction work continues. Many benefits have been obtained. From hydroelectricity generation alone, until the end of 1986, a total of 32.877 billion kWh of electrical energy, valued at 2.137 billion yuan, has been produced (see table 3). On the basis of the current speed of installation of generating capacity, it is estimated that by the

Table 3

Gezhouba Hydroelectric Plant: Electricity Produced and Investment by Year, 1981–1986

	Hydroelectricity generation			
Year	Installed capacity (10,000 kW)	Annual electrical energy output (100 million kWh)	Value of electrical production (100 million yuan)	Project investment (100 million yuan)
1981	16.5	2.25	0.15	0.8171
1982	71.5	32.79	2.13	3.7151
1983	96.5	60.88	3.96	3.6798
1984	96.5	70.84	4.60	3.4998
1985	96.5	72.88	4.74	3.9995
1986 (partial)	164.0	89.13	5.79	2.7000
Total		328.77	21.37	18.411

originally planned completion date of the project in 1989, 70 billion kWh will have been produced. This is unprecedented among the completed hydroelectrical plants in the country. The vast amount of electrical power produced by Gezhouba now guarantees Central China a regular supply of energy for its industrial and agricultural development and has eased the tense power supply situation. It can be inferred that the economic and social benefits as a consequence of this will be even more significant.

Although the benefits from navigation can only be fully realized after the completion of the TGP, Gezhouba has shown that even on its own it is capable of creating navigational improvements, for the reservoir has effectively inundated nineteen shoals and banks. Of these, such well-known danger spots as Qingtan, Xietan, and Konglingtan have been thoroughly improved. The channel above and below Badong is also improved. The per hp tugboat haulage in the reservoir has increased from 0.4–0.5 t to 1.3 t; per 1,000 t shipping cost has declined from 70–85 (in 1979) to 32–40 yuan. In addition, the project has endowed the Gezhouba Project Office with construction equipment whose original worth was more than 500 million yuan. The office now also owns many production facilities and living quarters

(242,400 m^2 of plant space and 1,202,300 m^2 of living space) and has a large construction team of 50,000 skilled workers. All these constitute a reliable base for the TGP. They are also part of the benefits of the project.

From the above analysis, it is clear that the design estimates for the Gezhouba are credible. The landform and geological conditions at the site of the TGP are far more favorable than those at Gezhouba. It is obvious that because of a larger difference in elevation and a more easily controlled flow, the TGP will be more cost effective than Gezhouba.

10

Assessing the Three Gorges Project from Three Macrostrategic Aspects

Tian Fang, Lin Fatang, and Ling Chunxi

THE THREE GORGES Project is a mega project that has captured worldwide attention. The project not only requires huge investment, extends over a long construction period, and touches on numerous problems and various complicated, interrelated factors, it also involves such areas as science and technology, economics, politics, society, environment, and national defense. The soundness of decisions made on the TGP will have a tremendous impact on the entire national economy. Because the TGP is an integrated national mega project that transcends departments, industries, and regions, considerations must not be based solely on the partial benefits generated for one department, one region, or in one particular period of time. Instead, assessment must be viewed from the long-term and overall benefits to the people. A systematic evaluation of the TGP should not be confined to the technical aspects such as hydrology, siltation, inundation, resettlement, geology, earthquakes, ecology, hydraulic structures, and electrical engineering, or benefits like power generation, flood control, and navigation. The project should be assessed from the following three macro aspects: (1) comprehensive development

Tian Fang, Lin Fatang, and Ling Chunxi, "Cong sange cengci de hongguan juece shang lunzheng sanxia gongcheng." *Qun yan* 9 (1987): 29–31. Translated by Alexandra Poon.

of the entire Changjiang River basin; (2) the overall energy policy in China; and (3) the overall strategy for economic and social development in China.

Even within the TGP itself, there are still many crucial issues that need to be studied in greater detail. First, the issue of siltation. Simulation experimentation is still underway. Decisions can only be made after "responsible conclusions and clear answers" are obtained. The problem of siltation of the Sanmen Gorge Project on the Huanghe River is a lesson that must be learned and remembered.

Second, the problem of inundation and resettlement in the reservoir area. The areal extent of the resettlement area and the scale of the cities, towns, and factories to be relocated are rare in the history of irrigation and hydropower construction in the world. When we speak of "resource-exploiting" migration, we refer to rebuilding new-type villages, towns, and cities according to social systems engineering. This type of resettlement is not mere relocation and compensation, as in the past. Some hydropower stations in the world cannot be constructed because the issue of resettlement cannot be resolved.

Third, the issue of key facilities and equipment. For example, at present there is no prior example of the design for and making of facilities for vessel passage (required by the TGP).

Fourth, the issue of geology, earthquakes, and the ecological environment. At present, landslides are common in the dam and reservoir areas of the Three Gorges. This problem will be aggravated after the reservoir is impounded, and there is a possibility of inducing earthquakes. These phenomena may trigger a chain reaction, leading to disastrous consequences. Moreover, because dams and big reservoirs will be built on the main channel and tributaries of the Changjiang River, detailed studies of the long-term effects on the ecological environment should also be carried out.

The TGP has multiple benefits in many areas such as power generation, flood control, and navigation. However, what is its prime economic objective? Answers have varied at different times. Sometimes it was said to be for power generation, sometimes for flood control. What is the prime objective? Even if there are multiple objectives, they should be listed by priority. In fact, there are other superior schemes that can fulfill the stated objectives of the proposed TGP.

On power generation, the TGP should be assessed from the perspective of the overall national energy plan. Evaluation of the project should be placed in the context of the entire nation's power generation. The first consideration is to estimate the demand for power if the target of quadrupling [the gross value of industrial and agricultural output] by the year 2000 is to be realized. Studies should be carried out to determine which type of power stations will require less capital investment and shorter construction periods and have fewer problems and better returns. The demand for electricity after the year 2000 will then be assessed. Only by comparison with other thermal and hydropower stations can the merits of the TGP be estimated. For example, within the service region of the TGP there are other hydropower projects on the Qingjiang River, Hanjiang River, and Duhe River in Hubei province; the Yuanshui River, Zishui River, and Lishui River in Hunan province; the Wujiang River in eastern Sichuan province and eastern Guizhou province; and the Oujiang River and Feiyun River in Zhejiang province. More thermal power stations can be built in East China, or electricity can be supplied to the region by thermal power stations to be built in the coal mining districts south of the Huaihe River. Building these hydro and thermal power stations requires less investment and shorter construction periods, yields quicker results, and better meets the needs of economic development. Generally speaking, following normal construction schedules and procedures, these power stations require only four to eight years to reach the stage of power generation, while the entire construction could take only five to ten years. These power stations may be constructed by stages and in groups. This will benefit the local areas and will be conducive to the balanced development of industrial and agricultural production in the country. This alternative is also favorable for capital accumulation and turnover. The rich reserve of hydropower on the main channel and tributaries in the upper reaches of the Changjiang River, upper reaches of Huanghe River, and Hongshui River can also be exploited by building a series of hydropower stations arranged in a staircase fashion. Losses caused by inundation of these reservoirs are less. The number of people to be displaced and the resettlement conditions are better. Moreover, the scale of construction is smaller but the overall economic returns are higher. If these hydropower stations are built first, not only would the electricity demand arising from quadrupling the industrial and agricultural output by the end of this century be met, the economic development of extensive areas of Southwest, Northwest, and South China, especially the minority nationality regions, can also be promoted. This will create favorable conditions for shifting the development priority to the west. Therefore, judging from the overall arrangement to meet the country's electricity needs, the proposition that the TGP should go ahead immediately cannot stand.

In terms of flood control, since flood waters originate from different tributaries in the middle and upper reaches of the Changjiang River, comprehensive studies of the entire river basin should be carried out. The TGP alone cannot solve the problem of flooding in the middle and lower reaches of the river. For example, the TGP contributes little to the flood control of Wuhan City. According to the analysis of the causes of flooding in different sections along the Changjiang River, a comprehensive flood control plan covering the main channel and tributaries in the upper and middle reaches of the river is necessary. Flood control measures should combine both retaining and discharging flood waters with emphasis on the latter. Comprehensive measures of integrating permanent with temporary solutions and engineering with nonengineering approaches should be adopted. To solve the problem of flooding on the Jingjiang section, engineering measures consisting mainly of heightening and reinforcing the Jingjiang embankments and dredging the river channel should be adopted. Flood diversion regions may have to be put to use when needs arise. In line with the comprehensive utilization of the river, reservoirs may gradually be constructed on the tributaries in the upper and middle reaches of the river. The TGP should not be viewed as the panacea for all the problems of flooding. Even when completed, the TGP can divert only a very small percentage of the total volume of the Changjiang flood waters due to the limited storage capacity of the reservoir. Effective flood control still depends on other integrated measures; the thought that "once the TGP is completed, everything will be fine" is unrealistic. This warning was emphasized in the document of the party's 1958 Central Committee meeting in Chengdu entitled "Comments on the Three Gorges Key Water Conservancy Project and on the Planning of the Changjiang River Basin."

On navigation, the point that "navigation is first" was reiterated by the late Premier Zhou Enlai. The Changjiang River is the main artery of water transportation across the eastern, central, and western regions of China. The upper, middle, and lower reaches of the river are one entity for navigation. If navigation on the Changjiang River is restricted by the Three Gorges Dam, then the economic development of the entire southwestern region, even the whole nation, would suffer. These losses are irreversible, and generation after generation of our descendants will have to bear the consequences. There are many methods to improve the waterway in the upper and middle reaches of the Changjiang River. Navigation on the Sichuan section of the river can be improved gradually year by year by means of river management. Moreover, during the construction period of the TGP, navigation along the river will be seriously hampered.

As the TGP is such a massive project, its evaluation must be placed in the context of China's overall economic and social development strategy.

1. The macroeconomic aspect. This refers to the scale of investment and financial capability of China. The estimated total investment of 20 billion yuan put forward by the Planning Office of the Changjiang River Basin (POCRB) obviously represents

wishful thinking. Other estimates computed by the departments and experts concerned, using different methods of calculation (interest paid during the construction period included), have provided such other figures as 45.9 billion, 60 billion, 70 billion, and 100 billion yuan. All these figures far exceed the estimates suggested by the POCRB. As the scales of the project and the methods and specifications of calculation vary in the above estimates, results were different. These discrepancies point to the need for more careful studies on the methods of calculation and on the scope of investment of the TGP. Failure to do so will result in a poor grip on the issue, and the consequences may be disastrous. As far as the economic benefits of the entire society are concerned, a dynamic, not a static, approach must be adopted. The most fundamental question is, however, the financial capability of China, that is, is the TGP affordable? Premier Zhao Ziyang has repeatedly warned in his reports on the Sixth and Seventh Five-Year Plans: "The scale of investment must be compatible with the financial capability of China. This is a decisive factor in ensuring a stable economy. This is also a macro principle of socialist economic construction. When the scale of investment is too large, serious imbalances will result, leading to setbacks in economic development and slowing down the progress of modernization." With reference to the TGP, this is the first, and also an extremely important, issue to be decided.

2. The TGP will not help realize the national goal of quadrupling the industrial and agricultural output by the end of this century. The TGP requires huge capital investment, and the construction period will extend for about two decades. Even if the project is initiated immediately, it can still fail to play a role in achieving the above national goal within this century. Moreover, should the construction of the TGP go ahead, construction of other large- and medium-size hydropower plants is bound to be adversely affected, resulting in the disruption of the overall plan of capital construction and the phasing of national development strategy. This will certainly slow down the pace of achieving the national economic goals by the end of this century.

3. The assessment of the TGP cannot be separated from the overall reform in China. Premier Zhao Ziyang emphasized in the "Report on the Seventh Five-Year Plan" that the task of reform is arduous, and he demanded that "all regions departments and enterprises must establish a holistic perspective. It is not permissible to gain one's own and partial interests at the expense of overall benefits. Self-interest may be pursued only when the overall benefits are also ensured." To concentrate all the resources on achieving this priority task, and in view of the present national financial conditions, construction of the TGP is not recommended, as it requires such a huge capital investment and a long construction period, and as it involves so many conflicts and unresolved problems. The impact of the project on China's overall reform must be studied carefully.

4. The issue of stability and unity. It has been widely recognized that the nine years after the Third Plenum of the party's Eleventh Central Committee can be regarded as the best period since 1949. To achieve such a stable and united situation has not been easy, and care should be exercised to consolidate and develop it further. In the latter phase of the Sixth Five-Year Plan period, however, China's economy was "overheated" due to excessive economic growth. Problems caused by the overexpansion of demand had become acute, and although some adjustments were made in 1986, demand still far exceeds supply, and the adverse balance of credits and fiscal and foreign exchange deficits remains relatively large. There are many problems in matching supply with demand, and it is difficult to curb price rises altogether. Nor is it easy to control the scale of investment. Many major raw materials are still scarce. All these show that the development of the economy must continuously be "compressed." If the TGP is to start hastily, additional tens of billions of yuan need to be spent. It is obviously difficult for the state to support this scale of investment. If forced to, it would inevitably aggravate the tension in the economy and be detrimental to the political stability and unity of the country. Issues as to whether the TGP should go ahead, and when, cannot be separated from con-

siderations of the economic and political situation of the country.

5. The issue of national defense. In any consideration of the TGP, the issue of national security cannot be overlooked. Although it is predicted that no major war will be fought within this century, what may happen in the twenty-first century is unpredictable. If anything happens, the TGP will become the first target of the enemy. It will affect the safety of tens of millions of people in the middle and lower reaches of the Changjiang River. This issue cannot be treated lightly.

In short, assessment of the TGP must be carried out on democratic and scientific bases, and opinions from experts and scholars in related fields throughout China should be gathered. Vigorous fact-finding studies have to be carried out in an environment of democracy, equality, and dialogue, so that different opinions can be gathered and decisions made carefully and rationally. Ignorance and wrong decisions based on incomplete information can be avoided. This is what the party and the people in China hope to achieve.

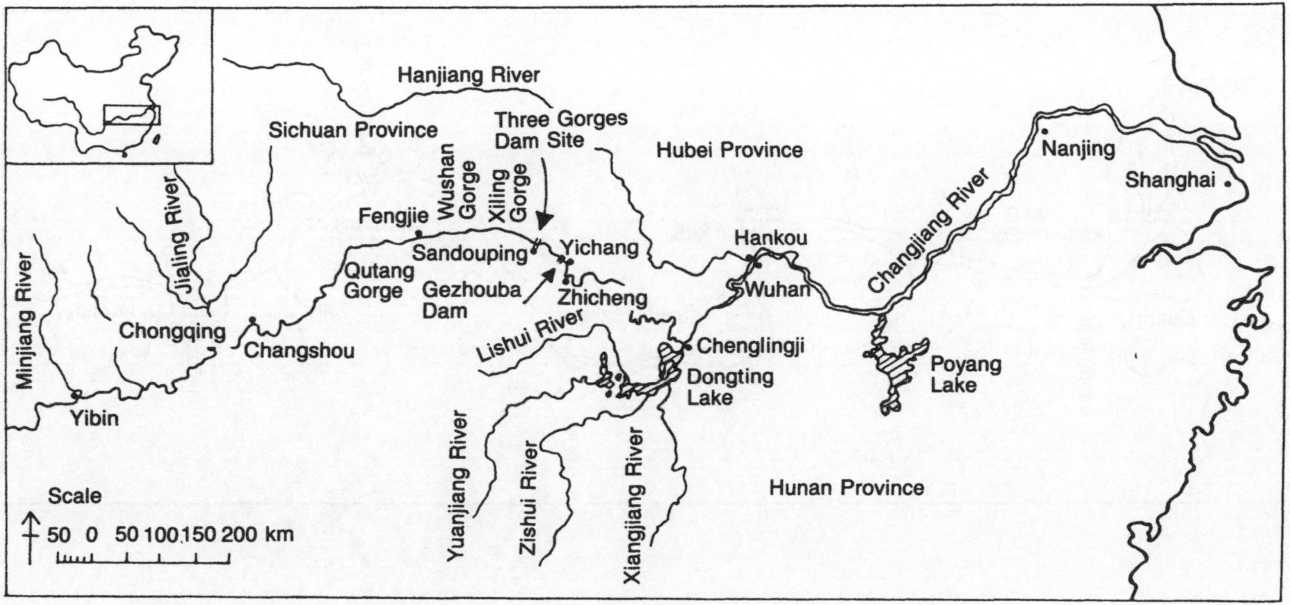

Appendix 1. The Changjiang River and the Three Gorges Project

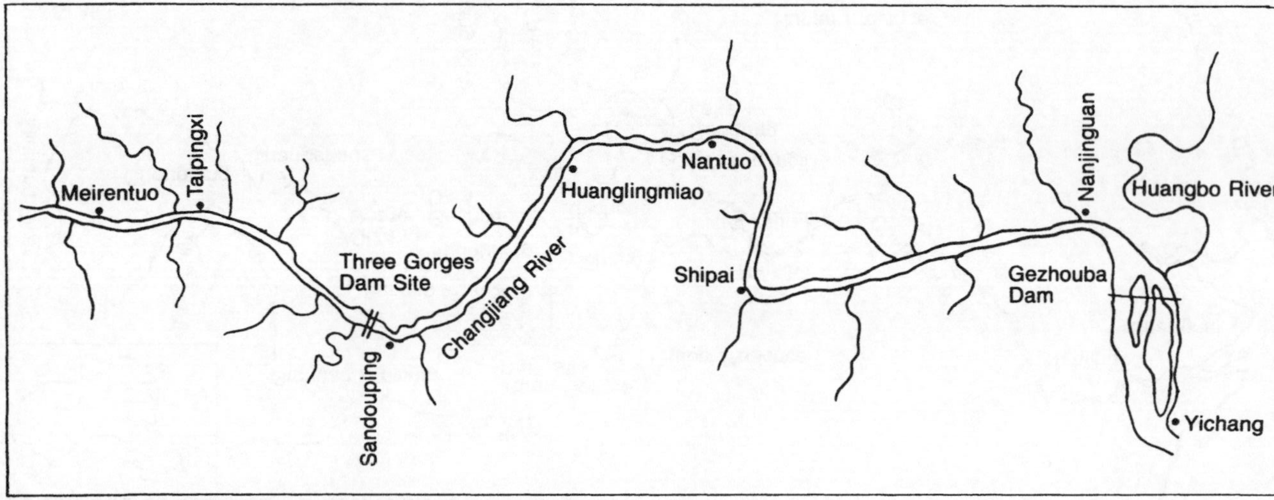

Appendix 2. Site of the Three Gorges Project

Appendix 3

Construction Data on the Three Gorges Dam

1. Layout

River width at dam site: 2,500 m
Probable maximum flood: 120,000 m^3/s
Concrete gravity dam length: 1,924 m
Maximum height above foundation: 165 m
Crest elevation: 175 m
Normal pool level: 150 m
Service spillway openings (23) elevation: 140 m, 8x15 m
Gated outlets (24) elevation: 85 m, 7x9 m
Emergency spillway (4 bays) elevation: 140 m, 10x15 m
Lock, double-line, 3-step: 280x34x5 m
  total lift: 87 m
Average lift each stage: 29 m
  capacity: 50 million tonnes/year
Vertical shiplift tank: 140x16x32 m
  capacity: 4.6 million tonnes/year
Powerhouse, left: 14x500 mega-watts
  right: 12x500 mega-watts

2. Major Summary of Construction Materials

Excavation: 52.4 million m^3
Concrete: 24 million m^3

J. F. Kennedy, "Sediment and Flood-Control Aspects of the Three Gorges Project, China," *Science & Technology Review*, no. 1 (1987): 13 (special issue: Symposium on the Three Gorges Project, December 6, 1986, New York).

Earth fill: 22 million m^3
Reinforcement steel: 207,000 t
Structure steel: 191,500 t

Appendix 4

Hydrologic and Flood Prevention Data on the Three Gorges and Changjiang River

1. Characteristics of the Changjiang River

Total length: 6,380 km; upstream of Yichang, 4,529 km
Drainage area: 1.8 million km^2; upstream of Yichang, 1 million km^2 (55% of total)
Average annual runoff volume: 995 billion m^3; upstream of Yichang, 452.9 billion m^3 (46% of total)
Mean annual water discharge: 31,500 m^3/s; upstream of Yichang, 14,300 m^3/s (45% of that of the entire river basin)
Mean annual sediment load: 478 million tonnes; upstream of Yichang, 521 million tonnes (including alluviation in the lake region)
Embankments along the main river channel and tributaries: 32,800 km, 3,748 km of which are major embankments on the main channel

---

Excerpts from Li Rui, *Lun sanxia gongcheng* (On the Three Gorges Project) (Hunan: Hunan kexue chubanshe, 1985), pp. 150–59. Translated by Ngan-chu Liu and W. X. Cheng.

2. Characteristics of the Upper Reaches and Tributaries of the Changjiang River

River	Length (km)	Drainage area (km²)	Mean average annual discharge (m³/s)	Peak flood discharge (m³/s)	Year of peak flood	Potentially exploitable water power: Installed generator capacity (10,000 kW)	Potentially exploitable water power: Annual power generation (100 million kWh)
Jinshajiang	3,500	502,000	4,640	29,000	1966	5,890	3,230
Yalongjiang (tributary)	1,187	145,200	1,430	8,680	1954	2,490	1,520
Minjiang	793	133,600	2,870	51,500	1917	3,060	1,670
Daduhe (tributary)	910	82,800	1,510	9,630	1939	2,510	1,370
Jialingjiang	1,119	159,800	2,130	53,600	1870	870	410
Wujiang	1,018	87,600	1,590	30,600	1830	830	420
Qingjiang	408	17,300	407	18,900	1909	170	90
Yuanshui	1,060	90,400	2,130	50,000	1766	350	156
Zishui	590	28,100	752	21,900	1926	120	62
Xiangshui	817	94,230	2,010	25,600	1926	100	47
Hanjiang	1,532	173,600	1,710	57,900	1935	610	250
Xiushui	350	14,700	258	13,000	1901	21	8
Ganjiang	744	81,700	2,030	26,100	1876	220	80
Fuhe	340	17,100	429	16,500	1876	22	8
Xinjiang	310	16,800	544	16,200	1878	18	7
Raohe	250	15,400	142	10,490	1884	22	7

3. Flood Discharge at Yichang

Year	Maximum daily discharge Rank	Maximum daily discharge m³/s	30-day flood flow Rank	30-day flood flow 100 million m³	60-day flood flow 100 million m³	Remarks
1870	1	110,000	1	1,767–1,852	—	820-year flood (based on historical survey)
1860	2	92,500	2	1,550	—	164-year flood (based on historical survey)
1788	3	86,000	3	1,450	—	137-year flood (based on historical survey)
1896	4	71,100	6	1,281	2,085	Highest peak discharge observed
1945	5	67,500	8	1,239	2,047	
1954	6	66,100	4	1,386	2,448	Largest volume of 30-day and 60-day flood flows observed
1921	7	64,800	7	1,278	2,022	
1931	8	64,600	17	1,065	1,893	
	⋮					
1949	26	57,900	11	1,133	1,979	More severe flooding in middle and lower reaches; not so severe at Yichang
	⋮					
1935	34	56,900	68	868	1,482	More severe flooding in part of Qingjiang, Lishui, and Hanjiang rivers
Mean annual flood		52,000		935	1,640	
Estimated 100-year flood		83,700		1,421	2,444	
Estimated 1,000-year flood		98,800		1,627	2,790	
Estimated 10,000 year flood		113,000		1,823	3,100	

Source: Based on empirical observations, 1877–1977; and historical records, 1153–1977.

4. Flood Level at Hankou (Wuhan Pass), 1865–1972

				Total flood flow volume of four selected months of the highest flow (100 million m^3)			
Year	Rank	Highest flood level (m)	Rank	Month	Flood flow in Hankou (1)	Flood flow in Yichang (2)	(2)/(1) × 100%
1954	1	29.73	1	5 - 8	6,102	3,367	55%
1931	2	28.28	2	7 - 10	5,150	3,132	61%
1935	3	27.58		7 - 10	4,736	2,846	60%
1968	4	27.39		7 - 10	4,657	3,286	71%
1870	5	27.36	4	6 - 9	4,870	—	—
1969	6	27.19		7 - 10	3,983	2,354	59%
1949	7	27.12	3	6 - 9	4,920	3,578	73%
	⋮	⋮					
1896	28	26.12		7 - 10	4,216	3,405	

Notes: According to field surveys, the most severe flood in Yichang occurred in 1870, but the flood level of Hankou in the same year was not high. Nor was it high in 1896, which was the year the maximum flood peak was observed in Yichang. In 1935, when a relatively low flood peak was observed in Yichang, Hankou was affected by the floodwater from the Hanjiang River. Flooding in these three years was all caused by localized rainstorms.

At the Hankou station, the 1954 flood was the highest on record. It was estimated to be a 100-year-flood. The 1931 flood was a relatively severe one.

Water level at which authorities should be alerted: 24.5 m; water level at which warning should be sounded: 26.3 m; previously estimated maximum water level: 29.73 m; estimated maximum water level according to 1972 scheme: 29.73 m.

5. Flood Level in Nanjing

Maximum water level: 10.22 m
Previously estimated maximum water level: 10.22 m
Estimated maximum water level according to 1972 scheme: 22.5 m

6. The 1954 Flood

According to records at the Hankou station, the 1954 flood was a severe flood that occurs once every one hundred years. The total

Location	May	June	July	August	May-August	% of total runoff volume of Balijiang
			(in 100 million m³)			
Yichang	391	479	1,171	1,326	3,367	42
Qingjiang, Juzhang, Xiangshui, Zishui, Yuanshui, and Lishui rivers, and areas in between	536	743	676	310	2,265	28
Lushui and Hanjiang rivers, e.g., Baxi and areas in between	111	141	345	221	818	10
Ganjiang, Fuhe, Xinjiang, Raohe, and Xiushui rivers, and areas in between	425	620	408	129	1,582	20
Aggregate volume of water flow upstream of Balijiang	1,462	1,983	2,600	1,987	8,032	100
Of which:						
Yuanshui River	178	193	258	145	403*	
Hanjiang River	42	32	154	196	350*	

Note: Natural lakes and river channels in the middle and lower reaches of the Changjiang River can hold about 147 billion m³ of water, while the capacity of flood diversion created by breaching embankments is about 100 billion m³. In 1954, the area of farmland flooded amounted to 475.5 million mu, including farmland affected by waterlogging, dike collapse, planned flood diversion, and mountain torrents.

*July-August only.

runoff volume at Balijiang (downstream of Poyang Lake) was the highest in May, June, July, and August.

7. The 1870 Flood

According to field surveys, the flood levels marked on bedrock between Chongqing and Wushan are relatively reliable. Though the peak discharge and the thirty-day flood flow thus estimated are not very accurate, it is certain that an especially severe flood occurred upstream of Yichang in 1870.

In 1870, a staff for stage height had already been installed by the Customs Office in Hankou. The peak flood level recorded in that year was 27.36 m, which ranks fifth in the annual maximum series. It is 2.37 m lower than the peak flood level of 1954. The total volume of flood flow for a four-month period was far less than in 1954.

It can be seen that floodwater in the upper reaches and that downstream did not converge in 1870. The hardest hit area after Songzikou was breached was the Dongting Lake area, which was flooded. Hankou and its downstream areas were not seriously affected by the flood.

Bibliography of Publications in English

Anonymous. "Kilowatts for the Next Century," *China News Analysis* 1378 (February 1, 1989): 1–10.

Anonymous. "Massive Chinese Hydro Project Poses Serious Ecological Risk," *International Wildlife* 15 (November-December 1985): 25.

Anonymous "Three Gorges Project: Main Points of the Feasibility Study Report," *Beijing Review* 32 (July 3, 1989): 22.

Anonymous. "Yangtze Project Study Approved," *Beijing Review* 12 (March 20, 1989):9–10.

Barbour, George. *Physiographic History of the Yangtze.* Peiping (Beijing): Peiping National Geological Survey of China and Institute of Geology, National Academy of Peiping, 1935.

———. "Physiographic History of the Yangtze," *Geographical Journal* 87 (1936): 17–34.

Bishop, Isabella Lucy Bird. *The Yangtze Valley and Beyond.* New York: G. P. Putnam, 1900.

Blakiston, Thomas Wright. *Five Months on the Yang-Tze.* London: J. Murray, 1862.

Boxer, Baruch. "China's Three Gorges Dam: Questions and Prospects," *The China Quarterly*, no. 113 (March 1988): 94–108.

Dixon, J. A., and K. W. Easter. "Economic Analysis at the Watershed Level," in *Watershed Resources Management—An Integrated Framework With Studies from Asia and the Pacific*, ed. K. W. Easter, J. A. Dixon, and M. M. Hufschmidt. Studies in Water Policy and Management, no. 10. Boulder: Westview Press, 1986.

Fearnside, Philip M. "China's Three Gorges Dam: Fatal Project or Step Toward Modernization?" *World Development* 16 (May 1988): 615–30.

Forestier, Katherine. "China Moves Towards a Damming Verdict," *New Scientist* 122 (April 1, 1989): 23.

Gaines, M. "China Reevaluates Three Gorges," *International Dams Newsletter* 2 (1987): 6.

Han Baocheng. "The Benefits of the Three Gorges Project," *Beijing Review* 29 (July 28, 1986): 22–25.

Han Baocheng. "Three Gorges Project: is it Feasible?" *Beijing Review* 29 (July 21, 1986): 16–20.

Jhaveri, Nayna. "The Three Gorges Debacle," *The Ecologist* 18, 2–3 (1988): 56–63.

LaBounty, J. F. "Assessment of the Environmental Effects of Constructing the Three Gorges Project on the Yangtze River," *Water International* 9 (1984): 10–17.

Lampton, David. "Water Politics," *The China Business Review*, (July–August 1983): 10–17.

Lieberthal, Kenneth, and Michel Oksenberg. *Policy Making in China: Leaders, Structures, and Processes.* Princeton: Princeton University Press, 1988.

Machetzki, R. "The Changjiang Province as a Key Economic Area." In *Development and Distribution in China*, ed. C. K. Leung and J. C. H. Chai, pp. 257–81. Hong Kong: Centre of Asian Studies, University of Hong Kong, 1985.

Plant, Cornell. *Glimpses of the Yangtze Gorges.* Shanghai: Kelly and Walsh, 1926.

Ryder, Gráinne. "China's Three Gorges Project: Whose Dam Business Is it?" *Cultural Survival Quarterly* 12 (1988): 17–19.

———. "Exposing the Secrets of the Three Gorges Dam," *World Rivers Review*, special focus insert (January 1989).

———. *Damming the Three Gorges: What Dam-Builders Don't Want You to Know. A Critique of the Three Gorges Water Control Project Feasibility Study*. Toronto: Probe International, 1990.

Stavins, R. "The Toulumne River: Preservation or Development," Environmental Defence Fund, mimeo.

Sun Yueqi. "Three Gorges Project: Why I Am Against the Project," *Beijing Review* 32 (July 3, 1989): 31–35.

Van Slyke, L. P. *Yangtze: Nature, History and the River*. Reading, MA: Addison-Wesley, 1988.

Xiang Rong. "On-site Report: Building a Dam on the Changjiang River," *Beijing Review* 12 (1979): 17.

Yao Jianguo. "Experts Okay Three Gorges Project," *Beijing Review* 31 (December 12, 1988): 9–10.

———. "Three Gorges Project: Dream and Reality," *Beijing Review* 32 (July 3, 1989): 19–22.

———. "Yangtze Project Study Approved," *Beijing Review* 32 (March 20, 1989): 13–14.

INDEX

Shiu-hung Luk is Associate Professor of Geography at the University of Toronto. He received his Ph.D. from the University of Alberta. His areas of interest are physical geography, soil erosion, and land management.

Joseph Whitney is Professor of Geography at the University of Toronto. He studied geography at the University of Chicago from which he received his Ph.D. He is interested in environmental geography, especially environmental impact assessment.